**High Energy Density
Lithium Batteries**

*Edited by
Katerina E. Aifantis,
Stephen A. Hackney,
and R. Vasant Kumar*

Related Titles

Ozawa, K. (ed.)

Lithium Ion Rechargeable Batteries

Materials, Technology, and New Applications

2009
ISBN: 978-3-527-31983-1

Vielstich, W. (ed.)

Handbook of Fuel Cells

Fundamentals, Technology, Applications, 6-Volume Set

2009
ISBN: 978-0-470-74151-1

Lui, H., Zhang, J. (eds.)

Electrocatalysis of Direct Methanol Fuel Cells

From Fundamentals to Applications

2009
ISBN: 978-3-527-32377-7

Mitsos, A., Barton, P. I. (eds.)

Microfabricated Power Generation Devices

Design and Technology

2009
ISBN: 978-3-527-32081-3

Garcia-Martinez, J. (ed.)

Nanotechnology for the Energy Challenge

2010
ISBN: 978-3-527-32401-9

Kolb, G.

Fuel Processing

for Fuel Cells

2008
ISBN: 978-3-527-31581-9

Eftekhari, A. (ed.)

Nanostructured Materials in Electrochemistry

2008
ISBN: 978-3-527-31876-6

Sundmacher, K., Kienle, A., Pesch, H. J., Berndt, J. F., Huppmann, G. (eds.)

Molten Carbonate Fuel Cells

Modeling, Analysis, Simulation, and Control

2007
ISBN: 978-3-527-31474-4

High Energy Density Lithium Batteries

Materials, Engineering, Applications

Edited by
Katerina E. Aifantis, Stephen A. Hackney,
and R. Vasant Kumar

WILEY-VCH Verlag GmbH & Co. KGaA

The Editors

Dr. Katerina E. Aifantis
Aristotle Universiry of Thessaloniki
Laboratory of Mechanics and Materials
P.O. Box 468
52124 Thessaloniki
Greece

and

Physics Department
Michigan Technological University
1400, Townsend Drive
Houghton, MI 49931
USA

Prof. Stephen A. Hackney
Materials Science and Metallurgy
Michigan Techological University
1400, Townsend Drive
Houghton, MI 49931
USA

Dr. R. Vasant Kumar
University of Cambridge
Department of Materials Science
and Metallurgy
Pembroke Street
Cambridge, CB2 3QZ
United Kingdom

Cover Illustration
The cover was designed by Katerina E. Aifantis
and was inspired by the Si micron-cones that are
fabricated by Dr. E. Strakakis' group at
IESL–Forth, Crete, Greece.

■ All books published by Wiley-VCH are carefully
produced. Nevertheless, authors, editors, and
publisher do not warrant the information contained
in these books, including this book, to be free of
errors. Readers are advised to keep in mind that
statements, data, illustrations, procedural details or
other items may inadvertently be inaccurate.

Library of Congress Card No.: applied for

British Library Cataloguing-in-Publication Data
A catalogue record for this book is available from
the British Library.

**Bibliographic information published by the
Deutsche Nationalbibliothek**
The Deutsche Nationalbibliothek lists this
publication in the Deutsche Nationalbibliografie;
detailed bibliographic data are available on the
Internet at http://dnb.d-nb.de.

© 2010 WILEY-VCH Verlag GmbH & Co. KGaA,
Weinheim

Cover Design Adam-Design, Weinheim
Typesetting Toppan Best-set Premedia Limited
Printing and Binding Strauss GmbH, Mörlenbach

Printed in the Federal Republic of Germany
Printed on acid-free paper

ISBN: 978-3-527-32407-1

Katerina E. Aifantis would like to dedicate this book to Father Symeon Krayio-poulos for his continuous support and guidance throughout her academic endeavors.

High Energy Density Lithium Batteries. Edited by Katerina E. Aifantis, Stephen A. Hackney, and R. Vasant Kumar
© 2010 WILEY-VCH Verlag GmbH & Co. KGaA, Weinheim
ISBN: 978-3-527-32407-1

Contents

High Energy Density Lithium Batteries. Edited by Katerina E. Aifantis, Stephen A. Hackney, and R. Vasant Kumar
© 2010 WILEY-VCH Verlag GmbH & Co. KGaA, Weinheim
ISBN: 978-3-527-32407-1

Preface

Modern society is characterized by a constant need for energy. In the past century, the majority of this energy has been supplied by fossil fuels. Fossil fuels are not only used in the thermal generation of electricity, but oil products are also relied upon for rapid transportation. The oil shock of 2008 and the concerns of possible global climate change have brought into question this reliance on petroleum and there has been a discernable shift towards developing technologies that can convert alternative energy sources, such as solar, wind and nuclear into electricity. It is just as important, however, to develop, at the same pace, technologies that can store this energy in a portable form. Portable energy allows for the increasing interconnectivity of people around the world through rapid communication and transportation. The portable energy afforded by high energy density batteries has not only made possible a variety of personal communication, entertainment and computational devices, but has found use in biomedical implantable devices such as pacemakers. Recently, this portable energy has also begun to influence battery applications in transportation. The improvement of battery science and technology is, therefore, a critical link in transitioning from fossil fuels to alternative energy sources. In particular, the emerging field of nanotechnology promises not only higher energy density batteries, but also rechargeable batteries with longer lifetimes.

It is this need for high density energy storage devices that prompted us to edit the present book. Although batteries are essential to everyone from a very young age, their explicit study is not custom during college education. The present book, therefore, starts out with an introductory chapter that familiarizes the reader with the basic electrochemical processes and properties of batteries. In continuing, Chapters 2 and 3 give a historic outline of the development of primary and secondary (rechargeable) batteries, where the highly preferred properties of Li batteries are illustrated. To further motivate the reader about the importance of continuous research on secondary Li chemistries, Chapter 4 describes the current and potential application of Li batteries, focusing on how they can be used for powering electric vehicles. The remaining chapters, therefore, elaborate on technological developments that are currently being undertaken for improving cathodes, anodes and electrolytes for rechargeable Li batteries; the common characteristic of all these components is their nanoscale structure. This book is, therefore, appropriate

High Energy Density Lithium Batteries. Edited by Katerina E. Aifantis, Stephen A. Hackney, and R. Vasant Kumar
© 2010 WILEY-VCH Verlag GmbH & Co. KGaA, Weinheim
ISBN: 978-3-527-32407-1

not only for advanced undergraduates and graduates, but also for battery developers (Chapters 4-8). Chapter 8 in particular is a collection of recent studies that are concerned with the limited theoretical works of the last decade that try to predict, using mechanics, the optimum materials chemistries for next generation anodes and cathodes. Such theoretical considerations, must be accounted for, in order to develop next-generation electrodes, as experiments and theory go hand in hand for obtaining the most efficient product. And despite the fact that plethora experimental studies are concerned with anode and cathode materials, theoretical works that try to predict their damage during charging and discharging, by employing known theoretical models, are well below twenty.

In fact, it was this attempt of interpreting damage and fracture in Li-electrodes that initiated the collaboration of Katerina Aifantis and Stephen Hackney, 9 years ago, when the latter posed this research problem to the former who was his undergraduate student at the time. They soon realized that their theoretical predictions were in agreement with experimental data and were in fact able to employ their design criteria for fabricating promising Sn/C nanostructured anodes. This urged them to make the importance of theoretical input known to a larger audience, through publication of a book, and that is why the present book concludes with a chapter devoted to theoretical issues of Li batteries. In searching for a third editor that could contribute by providing more complete information on the operation of electrochemical cells, they decided to ask R. Vasant Kumar, not only due to his expertise, but also due to his thorough web-based lecture notes.

In order to ensure continuity throughout the book we wrote or co-wrote the majority of the chapters ourselves and we constantly communicated with our authors, whom we would like to gratefully acknowledge for their thoroughness and promptness. In this connection we would like to thank Dr. Emmanuel Stratakis of the IESL-FORTH, Crete, Greece, for providing SEM images of his Si-microstructures on which Katerina E. Aifantis based the cover influenced by the paintings of her artist mother. KEA would also to thank the European Research Council (ERC Starting Grant 211166) for currently supporting her research on nanostructured materials, including Li-anodes, as well as her mentor: Professor E.C. Aifantis (father), and PhD advisors J.R. Willis & J.Th.M. De Hosson. The Editors would also like to acknowledge use of materials from the University of Cambridge web-based teaching and learning package on "Batteries".

In concluding this preface we would like to mention that the past two decades have not only seen considerable improvements in performance of the well-established types of secondary batteries but also witnessed the introduction of new types of batteries. Over the same period many new applications for such batteries have also taken place in wide ranging applications. This is by no means the end of the story. In the near future new chemistries may emerge within the secondary battery technology resulting in applications we have not even imagined at this stage.

Katerina E. Aifantis, Stephen A. Hackney, R. Vasant Kumar

List of Contributors

Katerina E. Aifantis
Aristotle University of
Thessaloniki
Laoratory of Mechanics and
Materials
P.O. Box 468
52124 Thessaloniki
Greece

Michigan Technological
University
Physics Department
1400, Townsend Drive
Houghton, MI 49931
USA

Martin L. Dunn
University of Colorado
Department of Mechanical
Engineering
Boulder, CO 80309
USA

Stephen A. Hackney
Michigan Technological
University
Department of Materials Science
and Engineering
Houghton, MI 49931
USA

Seok Kim
Pusan National University
Department of Chemical and
Biochemical Engineering
San 30, Jangjeon-dong,
Geumjeong-gu
Busan 609-735
South Korea

R. Vasant Kumar
University of Cambridge
Department of Materials Science
and Metallurgy
Cambridge, CB2 3QZ
UK

Kurt Maute
University of Colorado
Aerospace Engineering Science
Boulder, CO 80309
USA

Soo-Jin Park
Inha University
Department of Chemistry
253, Yonghyun-dong, Nam-gu
Incheon 402-751
South Korea

High Energy Density Lithium Batteries. Edited by Katerina E. Aifantis, Stephen A. Hackney, and
R. Vasant Kumar
© 2010 WILEY-VCH Verlag GmbH & Co. KGaA, Weinheim
ISBN: 978-3-527-32407-1

Thapanee Sarakonsri
Chiang Mai University
Department of Chemistry
Chiang Mai 50200
Thailand

Min-Kang Seo
Inha University
Department of Chemistry
253, Yonghyun-dong, Nam-gu
Incheon 402-751
South Korea

1
Introduction to Electrochemical Cells

R. Vasant Kumar and Thapanee Sarakonsri

1.1
What are Batteries?

The purpose of this chapter is to provide the basic knowledge on batteries, which will allow for their general understanding. Therefore, after defining their components and structure, an overview of the quantities that characterize these storage devices will be given.

Scientifically batteries are referred to as electrochemical or galvanic cells, due to the fact that they store electrical energy in the form of chemical energy and because the electrochemical reactions that take place are also termed galvanic. Galvanic reactions are thermodynamically favorable (the free energy difference, ΔG, is negative) and occur spontaneously when two materials of different positive standard reduction potentials are connected by an electronic load (meaning that a voltage is derived). The material with the lower positive standard reduction potential undergoes an oxidation reaction providing electrons by the external circuit to the material with the higher positive standard reduction potential, which in turn undergoes a reduction reaction. These half reactions occur concurrently and allow for the conversion of chemical energy to electrical energy by means of electron transfer through the external circuit. It follows that the material with the lower positive standard reduction potential is called the negative electrode or anode on discharge (since it provides electrons), while the material with the higher positive standard reduction is called the positive electrode or cathode on discharge (since it accepts electrons).

In addition to the electrodes, the two other constituents that are required for such reactions to take place are the electrolyte solution and the separator. The electrolyte is an ion conducting material, which can be in the form of an aqueous, molten, or solid solution, while the separator is a membrane that physically prevents a direct contact between the two electrodes and allows ions but not electrons to pass through; it therefore ensures electrical insulation for charge neutralization in both the anode and cathode once the reaction is completed. Two final parts required to complete a commercial galvanic cell are the terminals. They are necessary when applying the batteries to electrical appliances with specific holder

High Energy Density Lithium Batteries. Edited by Katerina E. Aifantis, Stephen A. Hackney, and R. Vasant Kumar
© 2010 WILEY-VCH Verlag GmbH & Co. KGaA, Weinheim
ISBN: 978-3-527-32407-1

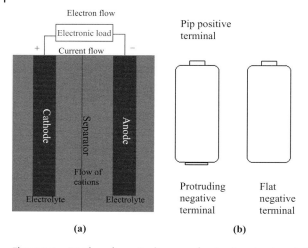

(a)   **(b)**

Figure 1.1 (a) The schematic diagram of a simple galvanic cell. (b) Terminal designs for cylindrical batteries.

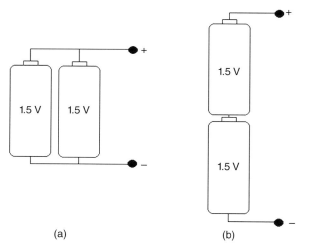

(a)   (b)

Figure 1.2 (a) An illustration of batteries connected in parallel to obtain double current. (b) An illustration of batteries connected in series to obtain 3 V.

designs in order to prevent short-circuit by battery reverse installation, and they are shaped so as to match the receptacle facilities provided in the appliances. For example, in cylindrical batteries, the negative terminal is either designed so as to be flat, or to protrude out of the battery end, while the positive terminal extends as a pip at the opposite end. A simple galvanic cell is illustrated in Figure 1.1a, while Figure 1.1b shows terminal designs for cylindrical batteries.

In order to meet the voltage or current used in specific appliances, cylindrical galvanic cells are connected in series or parallel. Figures 1.2a, and b represent

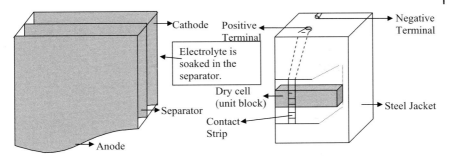

Figure 1.3 (a) Single-flat cell configuration; (b) composite flat cell configuration.

parallel and series connections; parallel connections allow for the current to be doubled, while series connections allow for the voltage to be doubled.

In addition to cylindrical battery cells, as those shown in Figures 1.1 and 1.2, flat battery configurations are also quite common. The biggest impetus for these configurations came from the rapid growth of portable radios, since the flat cells use the space of the battery box more efficiently than cylindrical cells. The electrodes are made in the form of flat plates, which are suspended in the electrolyte and are held immobilized in a microporous separator (Figure 1.3a). The separator also helps in isolating the electrodes, preventing any short-circuiting whereby ions can directly move internally between the anode and cathode. Short-circuiting will result in capacity loss, parasitic reactions, and heat generation. This can also lead to catastrophic situations causing fires, explosions, leakage of materials, and accidents. The configuration of Figure 1.3a can be scaled up to very large sizes, for high currents and large storage capacities, by placing each cell inside a plastic envelope and stacking them inside a steel jacket. Connector strips are used to collect and connect the positive and the negative electrodes to a common positive and negative terminal; a sketch of such a cell compaction is shown in Figure 1.3b.

Both cylindrical and flat cells come in various sizes so that they can fit a wide range of portable appliances and devices. Table 1.1 summarizes the various battery sizes that are available commercially.

1.2
Quantities Characterizing Batteries

Upon operation of galvanic cells, meaning that the device is on power mode, it is said that the galvanic cell is discharged and electrons flow, through an external circuit, from the anode to the cathode, which as a result attains a negative, and consequently cations are attracted from the anode to the cathode to which they diffuse through the electrolyte. The electrochemical reactions that take place upon operation of different batteries are shown in Table 1.2, whereas the quantities that characterize batteries are defined in Table 1.3.

Table 1.1 Dimensions of commercially available battery sizes [1].

Battery size	Diameter (mm)	Height (mm)
N	12	30.2
AAA	10.5	44.5
AA	14.5	50.5
C	26.2	50
D	34.2	61.5
F	32.0	91.0

Flat cells

Length (mm)	Width (mm)	Thickness (mm)
24	13.5	6.0
43	43	6.4

Rectangular cells

48.5	26.5	17.5

In order to better understand the differences between various materials chemistries, some of the quantities in Table 1.3 are further elaborated in the following pages.

1.2.1
Voltage

The theoretical standard cell voltage, E^0(cell) can be determined using the electrochemical series and is given by the difference between the standard electrode potential at the cathode, E^0(cathode), and the standard electrode potential at the anode, E^0(anode) [2] as

$$E^0(\text{cathode}) - E^0(\text{anode}) = E^0(\text{cell}) \tag{1.1}$$

The standard electrode potential, E^0, for an electrode reaction, written (by convention) as a reduction reaction (i.e., involving consumption of electrons), is the potential generated by that reaction under the condition that the reactants and the products are in their standard state in relation to a reference electrode. (A reactant or product is defined to be in its standard state when the component in a condensed phase is at unit activity and any component in the gas phase is at a partial pressure of 1 atmosphere.) In aqueous systems, the standard hydrogen potential is taken as the universal reference electrode, whose potential is defined as zero. In practical terms, the standard hydrogen electrode can be constructed by passing pure hydrogen at one atmosphere over an electrode of platinized platinum, where a high surface area of platinum is deposited on a platinum foil or plate, which is dipped into an acid solution of unit activity of H^+ ions, corresponding to 1 M acid solution. A list containing selected standard electrode potentials at 298 K in an aqueous

Table 1.2 Standard electrode potentials in an aqueous electrolyte at 298 K (written as reduction reactions by convention).

Reaction	E^0 (V)
$Li^+ + e^- \rightarrow Li$	−3.10
$Na^+ + e^- \rightarrow Na$	−2.71
$Mg^{2+} + 2e^- \rightarrow Mg$	−2.36
$\frac{1}{2}H_2 + e^- \rightarrow H^-$	−2.25
$Mn^{2+} + 2e^- \rightarrow Mn$	−1.18
$MnO_2 + 2H_2O + 4e^- \rightarrow Mn + 4OH^-$	−0.98
$2H_2O + 2e^- \rightarrow H_2 + 2OH^-$	−0.83
$Cd(OH)_2 + 2e^- \rightarrow Cd + 2OH^-$	−0.82
$Zn^{2+} + 2e^- \rightarrow Zn$	−0.76
$Ni(OH)_2 + 2e^- \rightarrow Ni + 2OH^-$	−0.72
$Fe^{2+} + 2e^- \rightarrow Fe$	−0.44
$Cd^{2+} + 2e^- \rightarrow Cd$	−0.40
$PbSO_4 + 2e^- \rightarrow Pb + SO_4^{2-}$	−0.35
$Ni^{2+} + 2e^- \rightarrow Ni$	−0.26
$MnO_2 + 2H_2O + 4e^- \rightarrow Mn(OH)_2 + 2OH^-$	−0.05
$2H^+ + 2e^- \rightarrow H_2$	0.00
$Cu^{2+} + e^- \rightarrow Cu^+$	+0.16
$Ag_2O + H_2O + 2e^- \rightarrow 2Ag + 2OH^-$	+0.34
$Cu^{2+} + 2e^- \rightarrow Cu$	+0.34
$O_2 + 2H_2O + 4e^- \rightarrow 4OH^-$	+0.40
$2NiOOH + 2H_2O + 2e^- \rightarrow 2Ni(OH)_2 + 2OH^-$	+0.48
$NiO_2 + 2H_2O + 2e^- \rightarrow Ni(OH)_2 + 2OH^-$	+0.49
$MnO_4^{2-} + 2H_2O + 2e^- \rightarrow MnO_2 + 4OH^-$	+0.62
$2AgO + H_2O + 2e^- \rightarrow Ag_2O + 2OH^-$	+0.64
$Fe^{3+} + e^- \rightarrow Fe^{2+}$	+0.77
$Hg^{2+} + 2e^- \rightarrow Hg$	+0.80
$Ag^+ + e^- \rightarrow Ag$	+0.80
$2Hg^{2+} + 2e^- \rightarrow Hg^+$	+0.91
$O_2 + 4H^+ + 4e^- \rightarrow 2H_2O$	+1.23
$ZnO + H_2O + 2e^- \rightarrow Zn + 2OH^-$	+1.26
$Cl_2 + 2e^- \rightarrow 2Cl^-$	+1.36
$PbO_2 + 4H^+ + 2e^- \rightarrow Pb^{2+} + 2H_2O$	+1.47
$PbO_2 + SO_4^{2-} + 4H^+ + 2e^- \rightarrow PbSO_4 + 2H_2O$	+1.70
$F_2 + 2e^- \rightarrow 2F^-$	+2.87

solution is given in Table 1.2. The batteries that make use of these materials as electrodes will be described in the next chapter. It should be noted that the standard electrode potential for a reaction defined arbitrarily as zero.

In order to obtain a true estimate of the actual open circuit cell voltage in the fully charged state for operation of the battery, the theoretical cell voltage is modified by the Nernst equation, which takes into account the nonstandard state of the reacting component as

$$E = E^0 - RT \ln Q \tag{1.2}$$

Table 1.3 Battery characteristics [2].

Battery characteristics	Definition	Unit
Open-circuit voltage	Maximum voltage in the charged state at zero current	Volt (V)
Current	Low currents are characterized by activation losses, while the maximum current is normally determined by mass transfer limitations	Ampere (A)
Energy density	The energy that can be derived per unit volume of the weight of the cell	Watt-hours per liter (Wh/dm^3)
Specific energy density	The energy that can be derived per unit weight of the cell (or sometimes per unit weight of the active electrode material)	Watt-hours per kilogram (Wh/kg)
Power density	The power that can be derived per unit weight of the cell	Watt per kilogram (W/kg)
Capacity	The theoretical capacity of a battery is the quantity of electricity involved in the electrochemical reaction	Ampere-hours per gram (Ah/g).
Shelf-life	The time a battery can be stored inactive before its capacity falls to 80%	Years
Service life	The time a battery can be used at various loads and temperatures	Hours (usually normalized for ampere per kilogram (A/kg) and ampere per liter $(A/lt^3))$
Cycle life	The number of discharge/charge cycles it can undergo before its capacity falls to 80%	Cycles

where $Q = a_{products}/a_{products}$ is the chemical quotient for the overall cell reaction and R is the gas constant ($R = 8.31$ J/(kmol)). Q is represented in the same way as the equilibrium constant K, except that the activities and partial pressures in Eq. (1.2) reflect the actual nonstandard values prevailing in the system. For example, for the electrode reaction

$$M^{2+} + 2\,e^- = M \tag{1.3}$$

the actual Nernstian electrode potential is

$$E = E^0 - RT \ln a_M/a_{M\,2+} \tag{1.4}$$

The Nernstian potential in Eq. (1.4) will change with time due to the self-discharge by which the activity (or concentration) of the electroactive component in the cell is modified. Thus, the nominal voltage is determined by the cell chemistry at any given point of time.

The operating voltage produced is further modified as a result of discharge reactions actually taking place and will always be lower than the theoretical voltage due to polarization (http://www.msm.cam.ac.uk/Teaching/mat1b/courseB/BH.pdf) and the resistance losses (IR drop) of the battery as the voltage is dependent on the current, I, drawn by an external load and the cell resistance, R, in the path of the current. Polarization rises in order to overcome any activation energy for the electrode reaction and/or concentration gradients near the electrode. These factors are dependent upon electrode kinetics and, thus, vary with temperature, state of charge, and with the age of the cell. Of course the actual voltage appearing at the terminal needs to be sufficient for the intended application.

1.2.2
Electrode Kinetics (Polarization and Cell Impedance)

Before continuing to the other quantities indicated in Table 1.3, the electrode kinetics, which as was previously shown affect the voltage, will be elaborated on. Thermodynamics expressed in terms of the electrode potentials can tell us the theoretical and open circuit cell voltage, as well as, how feasible it is for a cell reaction to occur. However, it is necessary to consider kinetics in order to obtain a better understanding of what the actual cell voltage may be, since the charge transfer and the rates of the reactions at the electrodes are usually the limiting factors. In continuing, therefore, the main kinetic issues that affect battery performance are summarized.

1.2.2.1 Electrical Double Layer
When a metal electrode is in an electrolyte, the charge on the metal will attract ions of opposite charge in the electrolyte, and the dipoles in the solvent will align. This forms a layer of charge in both the metal and the electrolyte, called the *electrical double layer*, as shown in Figure 1.4. The electrochemical reactions take place

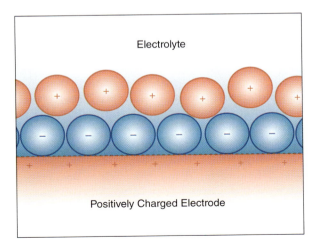

Figure 1.4 Illustration of double layer [2].

in this layer, and all atoms or ions that are reduced or oxidized must pass through this layer. Thus, the ability of ions to pass through this layer controls the kinetics, and is therefore the limiting factor in controlling the electrode reaction. The energy barrier toward the electrode reaction, described as the activation energy of the electrochemical reaction, lies across this double layer.

1.2.2.2 Rate of Reaction

The rates of the chemical reactions are governed by the Arrhenius relationship, such that the rate of reaction, k, is

$$k \propto \exp(-Q/RT) \tag{1.5}$$

where k is the activation energy for the reaction, T is the temperature in kelvin, and R is the universal gas constant.

In this case the rate of the reaction can be measured by the current produced, since current is the amount of charge produced per unit of time, and therefore proportional to the number of electrons produced per unit of time, that is, proportional to the rate of the reaction.

1.2.2.3 Electrodes Away from Equilibrium

When an electrode is not at equilibrium an *overpotential* exists, given by

$$\eta = E - E_0 \tag{1.6}$$

where η is the overpotential, E is the actual potential, and E_0 is the equilibrium potential.

Overpotential is used synonymously with polarization.

1.2.2.4 The Tafel Equation

The Tafel equation provides a relationship between the current and the overpotential during the oxidation or reduction reaction of an electrode. Consider a general reaction for the oxidation of a metal anode:

$$M \rightarrow M^{z+} + ze^- \tag{1.7}$$

The rate of this reaction, k_a, is governed by the Arrhenius relationship:

$$k_a = A \exp(-Q/RT) \tag{1.8}$$

where A is a frequency factor, which takes into account the rate of collision between the electroactive species and the electrode surface. From Faraday's law, one can express the rate in terms of the exchange current density at the anode, $i_{0,a}$:

$$i_{0,a} = zFk_a = zFK \exp(-Q/RT) \tag{1.9}$$

where $F = 96\,540\,C/mol$ is Faraday's constant. If an overpotential is now applied in the anodic direction, the activation energy of the reaction becomes

$$Q - \alpha zF\eta_a \tag{1.10}$$

where α is the "symmetry factor" of the electrical double layer, nominally 0.5.

Therefore the anodic current density, i_a, is

$$i_a = zFK \exp(-[Q - \alpha zF\eta_a]/RT) = zFK \exp(-Q/RT)\exp(\alpha zF\eta_a/RT) \quad (1.11)$$

which by Eq. (1.9) reduces to

$$i_a = i_{0,a} \exp(\alpha zF\eta_a/RT) \quad (1.12)$$

Equation (1.12) is known as the *Tafel equation*. By taking natural logs and rearranging, Eq. (1.12) can be written as

$$\eta_a = (RT/\alpha zF)\ln(i_a/i_{0,a}) \quad (1.13)$$

By setting $RT/(\alpha zF) = b_a$ and $\ln i_0 = -a_a/b_a$, Eq. (1.13) can be rewritten as

$$\eta_a = a_a + b_a \ln i_a \quad (1.14)$$

Or in terms of the anode potential, E_a,

$$\ln(i_a) = \ln(i_{0,a}) + (E_a - E_0)\alpha zF/RT \quad (1.15)$$

Solving Eq. (1.15) for E_a, gives

$$E = b_a \log(i_a/i_{0,a}) + a_a \quad (1.16)$$

where b_a is the anodic Tafel slope. Similarly, we can consider the reduction of metal ions at a cathode:

$$M^{z+} + ze^- \rightarrow M \quad (1.17)$$

The activation energy will be decreased by $(1-\alpha)zF\eta_c$, giving the cathodic current density as

$$i_c = i_{0,c} \exp([1-\alpha]zF\eta_c/RT) \quad (1.18)$$

and

$$\eta_c = (RT/([1-\alpha]zF)\ln(i_c/i_{0,c}) \quad (1.19)$$

Therefore, the cathode potential, E_c, is expressed as

$$E_c = b_c \log(i_c/i_{0,c}) + a_c \quad (1.20)$$

where b_c is the cathodic Tafel slope. A typical representation of a Tafel plot – a plot of log i vs E – is shown in Figure 1.5. Thus, for an applied potential, the current density, i, can be found from the Tafel plot in an electrolytic cell when the battery is being charged or discharged.

1.2.2.5 Example: Plotting a Tafel Curve for a Copper Electrode

Let us consider an electrode made of copper immersed in a half-cell containing copper ions at a 1 M concentration. The half-cell reaction for copper is

$$Cu^{2+} + 2e^- \rightarrow Cu \quad E^0 = +0.34 \text{ V} \quad (1.21)$$

The exchange current density for the above reaction is $i_0 = 1\,A\,m^{-2}$, reflecting the current density at zero overpotential, that is, at zero net reaction. Thus, the

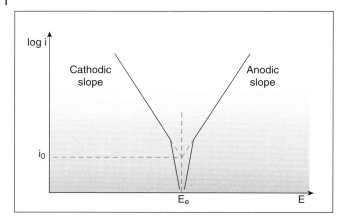

Figure 1.5 A typical Tafel plot [2], where i_0 is the exchange current density and E_e the equilibrium potential.

magnitude of the exchange current density is a reflection of the reversibility of a given electrode reaction.

For the Tafel equation

$$\eta = a + b\log(i) = a + b\log(i/i_0) \tag{1.22}$$

the general expression for the Tafel slope is

$$b = (\pm)2.303RT/\alpha zF \tag{1.23}$$

Taking $T = 300\,K$, and allowing for copper $\alpha = 0.5$ and $z = 2$, the Tafel slopes are calculated as $b_a = 0.059\,V/\text{decade}$ of current and $b_c = -0.059\,V/\text{decade}$ of current. Furthermore, for the anodic curve

$$\eta_a = E_a - E_0; E_a = E_0 + \eta_a \tag{1.24}$$

and for the cathodic curve

$$\eta_c = E_c - E_0; E_c = E_0 + \eta_c \tag{1.25}$$

The corresponding Tafel plot for copper is shown in the diagram of Figure 1.6. For example, during discharge, if the redox reaction is in the direction opposite of Eq. (1.21), where Cu is oxidized to copper ions into the solution, the electrode potential will be less than 0.34 V along the polarization line. The greater the operating current density, the lower the electrode potential. This in effect contributes to the reduction in the cell potential during discharge as a result of overpotential losses, signifying the energy barrier for the electron transfer reaction. On the other hand, during charging, the electrode potential increases with the applied current, thus, increasing the potential required for charging the cell back to its original state (by electrochemical reduction in this example). For a faster charging rate, a higher current density is desirable, but this can arise only at the expense of a higher applied voltage (higher energy) in order to overcome the increasing overpotentials.

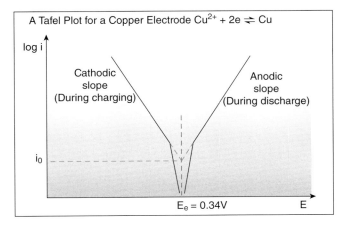

A Tafel Plot for a Copper Electrode $Cu^{2+} + 2e \rightleftharpoons Cu$

log i

Cathodic
slope
(During charging)

Anodic
slope
(During discharge)

i_0

$E_e = 0.34V$

E

Figure 1.6 Tafel plot for a copper electrode [2].

1.2.2.6 Other Limiting Factors

At very high currents a limiting current may be reached as a result of concentration overpotential, $\eta_{(conc)}$, restricting mass transfer rates to the diffusion rate of the electroactive species. A limiting current arises, which can be derived from Fick's first law of diffusion, under the condition that the electrode surface is depleted of the ion and the recovery of the ion concentration is limited by ion transport through the electrolyte diffusion boundary layer.

As the limiting current is diffusion limited, it can be determined by Fick's law of diffusion as

$$i_L = zFDC/\delta \tag{1.26}$$

where i_L is the limiting current density over a boundary layer, D is the diffusion coefficient of metal cations in the electrolyte, C is the concentration of metal cations in the bulk electrolyte, and δ is the thickness of the boundary layer. Typical values for Cu^{2+} for example would be $D = 2 \times 10^{-9} m^2 s^{-1}$, $C = 0.05 \times 10^4 kg\,m^{-3}$, $\delta = 6 \times 10^{-4} m$; these values give $i_L = 3.2 \times 10^2 A\,m^{-2}$.

The concentration of the overpotential, thus, represents the difference between the cell potential at the electrolyte concentration and the cell potential at the surface concentration because of depletion (or accumulation) at high current densities, given by

$$\eta_c(conc) = 2.303RT/zF \ln(i/i_L) \tag{1.27}$$

A Tafel curve showing this diffusion limiting of the current shown in Figure 1.7.

1.2.2.7 Tafel Curves for a Battery

In a battery there are two sets of Tafel curves present, one for each electrode material. During discharge one material will act as the anode and the other as the cathode. During charging the roles will be reversed. The actual potential difference

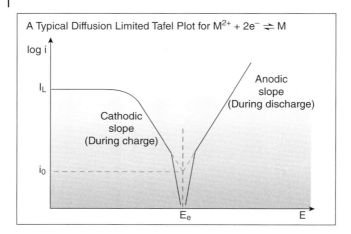

Figure 1.7 Diffusion limited current for the cathodic reaction [2].

between the two electrodes for a given current density can be found from the Tafel curve. The total cell potential is the difference between the anodic potential, E_a, and the cathodic potential, E_c.

In a galvanic cell, the actual potential, $V'_{cell,discharge}$, is less than the Nernst potential

$$V'_{cell,discharge} = E_c - |\eta_c| + E_\alpha - |\eta_a| \qquad (1.28)$$

Upon discharge the cell potential may be further deceased by the ohmic drop due to the internal resistance of the cell, r. Thus, the actual cell potential is given by

$$V_{cell,discharge} = V'_{cell,discharge} - iAr \qquad (1.29)$$

where A is the geometric area relevant to the internal resistance and i is the cell current density.

Similarly, on charging the applied potential is greater than the Nernstian potential, and can be calculated by the equation

$$V'_{cell,charge} = E_c + |\eta_c| + E_\alpha + |\eta_a| \qquad (1.30)$$

The cell charging potential may now be increased by the ohmic drop, and the final actual cell charging potential is given by

$$V_{cell,charge} = V'_{charge,harge} + iAr \qquad (1.31)$$

In summary, it can be stated that in order to maximize power density, it is important to achieve the most optimum value of cell potential at the lowest overpotentials and internal resistance. Usually at low current densities, overpotential losses arise from an activation energy barrier related to electron transfer reactions, while at a higher current density, the transport of ions becomes rate limiting giving rise to a current limit. Ohmic losses increase with increasing current, and can be further enhanced by the increased formation of insulating phases during the progress of

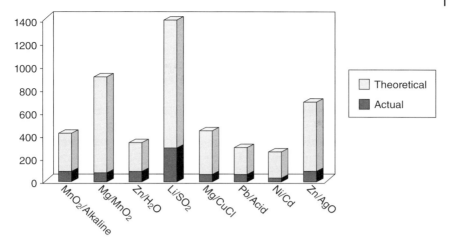

Figure 1.8 Theoretical and actual voltages of various battery systems [2].

charging. Power is a product of voltage and current; therefore, decreasing the current density by increasing the true surface area can also in principle result in a higher power density. However, unwanted side reactions may also be enhanced.

1.2.3
Capacity

The bar graph of Figure 1.8 shows the difference between the theoretical and actual capacities in $mAh\,g^{-1}$ for various battery systems.

The theoretical molar capacity of a battery is the quantity of electricity involved in the electrochemical reaction. It is denoted as Q_{charge} and is given by

$$Q_{charge} = xnF \tag{1.32}$$

where x is the number of moles of a chosen electroactive component that take place in the reaction, and n is the number of electrons transferred per mole of reaction. The mass of the electroactive component can be calculated as

$$M = xM_r \tag{1.33}$$

where M denotes the mass of the electroactive component in the cell and M_r the molecular mass of the same component. The capacity is conventionally expressed as $Ah\,kg^{-1}$ thus given in terms of mass, often called specific capacity, $C_{specific}$

$$C_{specific} = nF/M_r \tag{1.34}$$

If the specific capacity is multiplied by the mass of the electroactive component in the cell, one will obtain the rated capacity of a given cell. It is important to note that the mass may refer to the final battery mass including packaging or it may be reported with respect to the mass of the electroactive components alone. It is quite straightforward to recalculate the capacity in terms of the mass of the cell by dividing the rated capacity with the total mass of the cell.

In practice, the full battery capacity could never be realized, as there is a significant mass contribution from nonreactive components such as binders & conducting particles, separators & electrolytes and current collectors and substrates, as well as packaging. Additionally, the chemical reactions cannot be carried out to completion; either due to unavailability of reactive components, inaccessibility of active materials or poor reactivity at the electrode/electrolyte interface. The capacity is strongly dependent upon the load and can decrease rapidly at high drain rates as defined by the magnitude of current drawn, due to increased overpotential losses and ohmic losses which can exacerbate the problems with completion of the reaction. At higher drain rates denoting high operating currents, a battery will be discharged faster.

1.2.4
Shelf-Life

A cell may be subject to self discharge in addition to discharge during operation. Self-discharge is caused by parasitic reactions, such as corrosion, that occur even when the cell is not in use. Thus, the chemical energy may slowly decrease with time. Further energy loss may arise as a result of discharge during which insulating products may form or the electrolyte may be consumed. Therefore, shelf-life is limited by factors related to both nonuse and normal usage.

1.2.5
Discharge Curve/Cycle Life

The discharge curve is a plot of the voltage against the percentage of the capacity discharged. A flat discharge curve is desirable as this means that the voltage remains constant as the battery is used up. Some discharge curves are illustrated in Figure 1.9, where the potential is plotted against time as the battery is

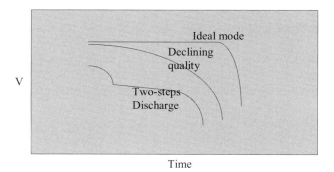

Figure 1.9 Change of voltage with time behavior in different cells [2].

discharged through a fixed load. In the ideal mode, the cell potential remains steady with time until the capacity is fully exhausted at the same steady rate and then it falls off to a low level. Some of the primary lithium cells display this type of nearly ideal flat discharge characteristics. In most other real batteries, the voltage may slope down gently with time as in primary alkaline cells or do so in two or more stages during discharge as in Lechlance cells.

1.2.6
Energy Density

The energy density is the energy that can be derived per unit volume of the cell and is often quoted as Wh/lt, where lt stands for liter. This value is dependent upon the density of the components and the design by which the various materials are interfaced together. In many applications availability of space for placing a battery must be minimized and thus, the energy density should be as high as possible without greatly increasing the weight of the battery to attain a given energy level. The battery flat cell, described in Figure 1.2, is an example of efficient designing that increases the energy density.

1.2.7
Specific Energy Density

The specific energy density, Wh/kg, is the energy that can be derived per unit mass of the cell (or sometimes per unit mass of the active electrode material). It is the product of the specific capacity and the operating voltage in one full discharge cycle. Both the current and the voltage may vary within a discharge cycle and, therefore, the specific energy, E, derived is calculated by integrating the product of the current and the voltage over time

$$E = \int V \cdot I \, \mathrm{d}t \tag{1.35}$$

The discharge time is related to the maximum and minimum voltage thresholds and is dependent upon the state of availability of the active materials and/or the avoidance of an irreversible state for a rechargeable battery. The maximum voltage threshold may be related to an irreversible drop of voltage in the first cycle, after which that part of the cycle is not available. The minimum threshold voltage may be determined by a lower limit below which the voltage is deemed to be too low for practical use or sets the limit for some irreversible losses, such that the system can only inadequately provide energy and power.

An active component may be less available due to side reactions, such as: (i) zinc reacting with the electrolyte in alkaline or silver oxide–zinc batteries, (ii) dendrite formation in rechargeable batteries, (iii) formation of passivation layers on the active components. Since batteries are used mainly as energy storage devices, the amount of energy (Wh) per unit mass (kg) is the most important property quoted for a battery. It must be noted that the quoted values only apply for the typical rates at which a particular type of battery is discharged. The specific

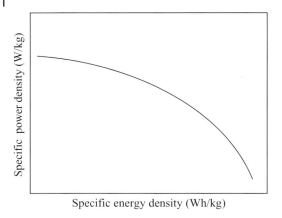

Figure 1.10 Ideal Ragone plot.

energy density values vary between 45 and 300 Wh/kg for primary batteries and 30 and 240 Wh/kg for secondary (rechargeable) batteries.

1.2.8
Power Density

The power density is the power that can be derived per unit mass of the cell (W/kg). At higher drains, signifying higher currents relating to higher power densities, the specific energy tends to fall off rapidly, hence, decreasing the capacity. This trade-off between power and energy density is best expressed in a Ragone plot, an idealized version of which is given in Figure 1.10. It is obvious that a certain battery has a range of values for specific energy and power, rather than a battery having a specific value of energy and power. In order to derive the maximum amount of energy, the current or the power drain must be at the lowest practical level. For a given cell chemistry, increasing the surface area of the electrodes can increase the cell's current at a given current density and, thus, deliver more power. The most efficient way to deliver a higher power density is to increase the effective surface area of an electrode while keeping the nominal area constant. It is important to consider any increase in parasitic reactions that may be enhanced due to the increase in the effective surface area. For example, in systems where corrosion is a concern, simply increasing the surface area may enhance the corrosion reactions while depleting the active material. Under these circumstances, the cell capacity will decrease along with the shelf-life.

1.2.9
Service Life/Temperature Dependence

The rate of the reaction in the cell is temperature dependent according to kinetics theories. The internal resistance also varies with temperature; low temperatures

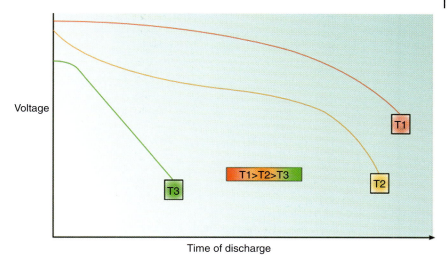

Figure 1.11 Effects of temperature on battery capacity [2].

give a higher internal resistance. At very low temperatures the electrolyte may freeze giving a lower voltage as ion movement is impeded. At very high tempera-tures the chemicals may decompose, or there may be enough energy available to activate unwanted, reversible reactions, reducing the capacity. The rate of decrease of voltage with increasing discharge will also be higher at lower temperatures, as will the capacity; this is illustrated in Figure 1.11.

1.3
Primary and Secondary Batteries

It should be mentioned that there are two main types of batteries: primary and secondary batteries. In primary batteries the chemical energy stored in the cell is such that it can be used only once to generate electricity, that is, once the cell is fully discharged it cannot be of further use. In secondary batteries the reverse redox reaction (also referred to as electrolysis and charging) can occur when the current is applied at a potential higher than the cell potential (E_{cell}) and the battery can be used reversibly many times. During charging, electrons flow to the anode through the external circuit and cations from the cathode diffuse through the electrolyte to the anode.

In Table 1.4, a timeline depicts the historic development of batteries separated into two blocks of primary and secondary battery histories. Presently, the most advanced technology of primary batteries is the Oxyride battery developed by Panasonic Corporation but not yet widely used. For the secondary batteries, the lithium polymer battery is the most advanced, mostly used as a power backup in

Table 1.4 History of electrochemical cell development tabulated with years and inventors.

Type	Year	Inventor	Battery
Primary batteries	1800	Alessandro Volta	Voltaic pile
	1836	John Frederic Daniel	Daniel cell
	1844	William Robert Grove	Grove cell
	1860	Callaud	Gravity cell
	1866	Georges-Lionel Leclanché	Leclanché wet cell
	1888	Carl Gassner	Zinc–carbon dry cell
	1955	Lewis Urry	Alkaline battery
	1970	No information	Zinc-air battery
	1975	Sanyo Electric Co	Lithium–manganese cell
	2004	Panasonic Corporation	Oxyride battery
Secondary batteries	1859	Raymond Gaston Planté	Planté lead-acid cell
	1881	Camille Alphonse Faure	Improved lead-acid cell
	1899	Waldmar Jungner	Nickel–cadmium cell
	1899	Waldmar Jungner	Nickel–iron cell
	1946	Union Carbide Company	Alkaline manganese secondary cell
	1970	Exxon laboratory	Lithium–titanium disulfide
	1980	Moli Energy	Lithium–molybdenum disulfide
	1990	Samsung	Nickel–metal hydride
	1991	Sony	Lithium-ion
	1999	Sony	Lithium polymer

laptop computers and slim and light weight mobile phones. The most common secondary battery used in communication devices, such as cellular phones, is the lithium-ion battery. Batteries are also sometimes classified in terms of the mode in which they are used:

- **Portable batteries:** These cover a wide range of batteries from those used in toys to those used in mobile phones and laptops.

- **Transport batteries:** The largest application of these is in the starting, lighting, and ignition (SLI) for cars or in electrical vehicles (e.g., in e-scooters and hybrid electrical vehicles).

- **Stationary batteries:** Include applications for stand-by power, backup in computers, telecommunications, emergency lighting, and for load-leveling with renewable energy, such as solar cells, during darkness, and wind during very calm weather.

Thus, batteries are correctly perceived as a critical enabling technology and future improvements are continuously sought. The various battery chemistries will be examined in detail in the following chapters.

1.4
Battery Market

It is estimated that the total market for all batteries in the year 2007 was worth $80 billion, with an annual growth of over 8%, and rapid growth in demand coming from China, India, Europe, South America, and Russia. Forecast for 2010 is shown in Table 1.5. It can be seen that the secondary batteries are expected to dominate over the primary batteries reversing the historical situation until recently. More than 50% of the primary market is from the alkaline cells, while more than half of the secondary market is dominated by lead-acid batteries.

In the secondary battery market, several new cells have been introduced just within the last two decades, for example, the Ni–MH cell (1990), Li-ion cell

Table 1.5 Global battery market forecast for 2010.

Type of battery	Global demand in US$, 98 billion
Primary	Primary total: 42%
Carbon–zinc	8%
Alkaline	22%
Others	12%
Secondary	Secondary total: 54%
Lead acid	28%
Ni–Cd/Ni–MH and others	12%
Li	14%

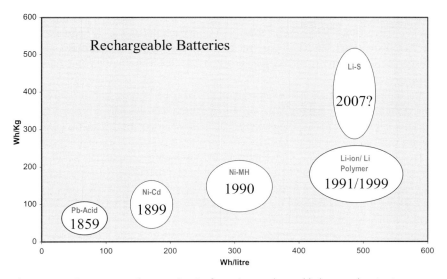

Figure 1.12 Comparison of energy density for various rechargeable battery chemistries.

(1991), rechargeable alkaline cells (1992), Li polymer (1999), and also the concept of mechanically rechargeable Zn-air batteries (2001). The new batteries, especially those based on the Li chemistries have not only met some of the existing demand, but can be said to have revolutionized the battery market by accelerating demand for laptops, mobile phones, and cordless hand held tool products.

An inspection of Figure 1.12 clearly reveals that with time, the specific energy (Wh/kg) and the energy density (Wh/lt) have continued to grow as batteries advance. It is this combination of high values of Wh/kg and Wh/lt that have been the key factors heralding this rapid growth. The number of batteries produced each year is staggering. The relatively newcomers, Li-ion batteries, alone are produced in quantities exceeding 1.2 billion cells per year.

1.5
Recycling and Safety Issues

The batteries consumed in households amount to approximately 4 billion cells per year. Batteries are important for everyday life but they can be a source of health problems and contamination. Besides the caution of using each type of battery the right battery for the appropriate function, the proper disposal is also a critical issue that needs to be urgently considered.

The alkaline cells do not have any toxic constituents, and although their disposal will incur energy and landfill penalty, they do not contaminate the environment with toxic pollutants. Some households have recharged primary alkaline cells despite warnings and against the advice of the manufacturers. Recharging is to some extent possible, provided the cells have not been discharged beyond 50% of their total capacity. With each recharge, the capacity of the cells begins to fade, further extenuated by the degree of depth of discharge. Furthermore, there is danger of explosion in recharging primary alkaline cells due to the release of hydrogen gas.

Rechargeable versions of alkaline batteries are made safe, but do suffer from low current capacity arising from the high internal cell resistance. In 1992, these types of cells were specifically developed as a low-cost power source for consumer goods.

It is important to note that choosing to purchase rechargeable batteries will decrease numerous global battery wastes. It is a fact that 600 primary batteries can be replaced by only two AA nickel cadmium secondary batteries. By new technology, rechargeable batteries come out as USB rechargeable batteries (see Figure 1.13), which is convenient and does not require a charger. However, due to the hazardous nature of heavy metals contained in a battery, some batteries have been reformulated. Alkaline batteries are manufactured as a mercury-free battery since 1992. The mercuric oxide button cell was finally replaced by mercury-free batteries, such as silver oxide or zinc air batteries. However, the nickel cadmium battery has not been totally replaced by other nonhazardous batteries yet. Moreover, some

Figure 1.13   USB rechargeable batteries (www.reghardware.co.uk).

appliances have the nickel cadmium battery built-in, which is difficult for battery recycling.

The annual number of batteries discarded run in billions worldwide; safe disposal and recycling are crucial in this important energy storage chain. Many countries are forced by legislations to have battery recycling plants such as in the USA, the UK, Germany, France, Turkey, Italy, and Poland. In the USA, the battery manufactures (Panasonic, Duracell, and Eveready Energizer) have formed a "Rechargeable Battery Recycling Corporations (RBRC) Battery Recycling Program" to take back dead batteries, especially button and Ni–Cd batteries. In the state of California all types of batteries are recycled. Used batteries can be disposed at any drop off site or at any participating retailers. Batteries are classified by the US federal government into nonhazardous and hazardous wastes. Used alkaline, carbon zinc, lithium-ion, nickel metal hydride, and rechargeable alkali manganese batteries are considered as nonhazardous waste and can be thrown away in the trash. Used Ni–Cd, lead-acid, silver oxide, and button cells containing mercuric oxide, silver oxide, lithium, alkaline, and zinc-air are classified as hazardous wastes and need to be recycled. Nonhazardous batteries must also be recycled, however, since throwing these batteries together with other wastes is dangerous as some of them may not be completely dead and may generate charge and spark which can lead to a fire and an explosion. They also should not be burned or incinerated which can result in an explosion and contamination of the environment.

In the near future, not recycling batteries will not be an available option. The lead-acid batteries have led the way in recycling, achieving over 95% recycling rates in many parts of the world such as in North America, Western Europe, and Japan. Emerging markets are also witnessing improving trends in recycling lead-acid batteries. Lead is relatively easy to recycle and recover as metallic lead. Currently, lead-acid batteries are recycled mainly using a high-temperature, pyrometallurgical process. A 10 000 ton per annum plant would be deemed small, with some plants processing 100 000 tons or even 200 000 tons of secondary lead per year.

In particular, the lead-acid battery recycling process stages are as follows:

- **Collection:** At repair workshops, garages, decontamination centers, clean points. The batteries and other lead-containing scrap are consolidated at intermediate agents or scrap dealers.

- **Transportation to appropriate plants.**

- **Materials preparation and sorting.**

- **Transportation to appropriate plants.**

- **Breakage:** Lead-acid batteries are composed of a variety of materials (Table 1.6), so they must be taken apart. Scrap batteries are broken apart in a hammer mill. The plastic casing (typically made of polypropylene, ebonite, and/or PVC) is shredded and recycled at extrusion plants. The sulfuric acid is also recovered and used in processes such as the manufacture of gypsum. Meanwhile, paste comprising lead sulfate ($PbSO_4$), and the lead metallic grids are extracted. In conventional lead recycling, the grids are normally added to the smelter; however, this is not always necessary, and instead they can be melted and refined at lower temperatures. Table 1.7 shows the typical composition of the lead-acid paste.

- **Neutralization:** Sodium hydroxide is added to the discharged battery paste to neutralize residual sulfuric acid and desulfurise, which avoids sulfur dioxide (SO_2) emissions during smelting. Aqueous sodium sulfate, possibly containing

Table 1.6 Typical composition of lead-acid battery [3].

Component	Wt.%
Lead–antimony alloy components (i.e., grids, poles, etc.)	25–30
Electrode paste	35–45
Sulfuric acid	10–15
Polypropylene	4–8
Other plastics (e.g., PVC, PE, etc.)	2–7
Ebonite	1–3
Other (e.g., glass, etc.)	<0.5

Table 1.7 Typical composition of lead-acid battery paste.

Material	Wt.%
Lead sufate	55–65
Lead dioxide	15–40
Lead monoxide	5–25
Metallic lead	1–5
Carbon black, plastics, fibers, other sulfates	1–4

some residual dissolved lead(II) together with colloidal particles, is then discharged. Depending on the country, the consent concentrations for this discharge to sewers range from a few ppm down to zero; the upper limits are likely to be decreased in the future.

- **Smelting or electrowinning:** The dry neutralized paste is usually smelted, along with the metallic grids, with a reducing agent in a high temperature Isasmelt or reverbatory or rotary furnace to produce lead. Coal or coke is the source of energy. The smelting takes place either in stand-alone reactors, or more often with lead concentrates derived from lead sulfide ores. Any lead or antimony slag is reheated with refining dross in a rotary furnace producing lead–antimony (Pb/Sb) alloy and waste slag. Instead of smelting, some large-scale operations dissolve the lead sludge/filter cake in powerful acids such as HCl, H_2SiF_6 or HBF_4 and recover lead by electrowinning.

- **Refining and blending:** A finished alloy is produced by adding the Pb/Sb alloy to the almost-pure lead metal in a refining and blending process.

- **Transport:** Assuming recycling and manufacture plants are separate entities, the lead then must be transported from the former to the latter.

- **New battery manufacture:** Recovered lead is reoxidized to PbO granules. Sulfuric acid (H_2SO_4) is added to the granules to form a paste applied to lead grids to manufacture new battery electrodes. After battery assembly, electrical energy is used to convert PbO to Pb in the anode and PbO_2 in the cathode. An overview of the lead-acid battery recycling process is given in Figure 1.14.

Over 90% of the lead from batteries can be extracted and reused, so with such high recovery rates, the secondary lead production is now a huge business. According to official estimates, more than 40% of the world's refined lead is now derived from secondary battery production. Table 1.8 presents some recycling input ratios (RIR) around the world. RIR quantify the contribution that recovered lead makes to the overall production of refined metal.

Lead-acid batteries are, in fact, one of the most efficiently recycled products in the world—albeit with varying environmental standards. In developed nations a variety of organized schemes are in operation. For example, Sweden, Germany, and Italy operate levy systems related to the lead market: when the lead price is so low that battery recovery is not economic, a levy is imposed on new batteries to finance recycling of used batteries. Italy and Ireland also have a national collection and recycling schemes. In the United States, many states require retailers to accept used car batteries when customers purchase new batteries. Several American states even require a cash deposit on new battery purchases, which is refunded to the consumer once the used battery is returned to the retailer. Elsewhere, battery collection is normally market driven. In the UK, for example, where organized systems do not exist, car repair shops and scrap metal dealers collect used batteries.

These RIR figures are based on official data collected from governments around the world. As such they are likely to underestimate true lead recovery rates because

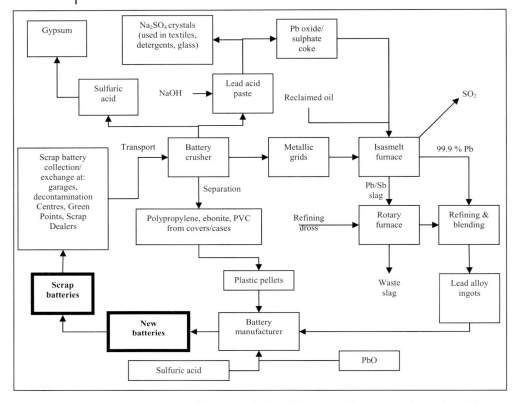

Figure 1.14 Overview of the current lead-acid battery recycling process flow (Adapted from: Brandon *et al.* 2003; Exide/GNB Technologies Poster ["*Recycling for a Better Environment*"]).

Table 1.8 Apparent recycling input ratios (RIR) for lead in 2006.

Region	RIR
Europe	62.7%
Americas	74.2%
Asia	15.7%
World	41.6%

in many countries, especially in the developing world, the informal nature of lead recycling makes obtaining reliable data on secondary lead usage impossible. This is particularly true for Asian countries: lead is so highly priced there that it is likely that almost 100% of the metal from batteries is reused. Since formal data may not be fully available, the details are thus anecdotal.

The recycling record in other types of batteries has still a long way to go to catch up with what is achieved for the lead-acid batteries. All spent batteries represent a valuable resource of metals and materials, such as Ni, Cd, Ag, Co, polymers, carbon, and oxides. New legislations are being steadily introduced to encourage recycling of batteries. For example, in 2006 Directives on batteries and accumulators were adopted in the European Union. The Directive has set new national targets for nations within the EU for collection of consumer batteries. For example, according to this Directive, the UK is required to collect 25% by weight of batteries sold by 2012 and the target is set to increase to 46% by 2016. New infrastructures have to be made available that allow consumer batteries to be safely collected, sorted, and recycled.

Some countries are leading the way for recycling consumer batteries, such as Belgium which has already achieved a collection rate that is over 65%. Disposal of some batteries such as Ni–Cd can cause major environmental problems. Cd is toxic and can seep into the water supply. Ni is also semitoxic. Primary Li batteries are fire hazards, especially if they are not fully discharged. Most secondary Li batteries contain the toxic element Co. Both primary and the secondary Li batteries contain toxic and flammable electrolytes. It must be stated, however, that, in general, recovery of metals from batteries is energy intensive. Therefore is pressing need for further research for developing environmentally friendlier recycling technologies that do not give rise to an unsustainable energy penalty.

Having given a general overview of batteries, starting from their main components, and configuration, to how they function and are disposed of, specific battery systems can now be examined. The first two chapters that follow present a historic timeline describing the development of batteries from the first galvanic cells of the 1800s to the commercial cells of the 21st century. It will be seen that Li batteries are the most promising high energy storage devices and therefore Chapters 4–8 focus on the next-generation Li cathodes, anodes, and electrolytes that will significantly increase their lifetime and range of applications.

References

1 Linden, D., and Reddy, T.B. (2002) *Handbook of Batteries*, 3rd edn, McGraw-Hill, New York.

2 University of Cambridge (2005) DoIT-PoMS Teaching and Learning Packages. University of Cambridge, Cambridge, http://www.doitpoms.ac.uk/tlplib/

batteries/index.php (accessed 5 February 2010).

3 EC (2001) Integrated Pollution Prevention and Control (IPPC). Reference Document on Best Available Techniques (BAT) in the Non Ferrous Metals Industries.

2
Primary Batteries

Thapanee Sarakonsri and R. Vasant Kumar

2.1
Introduction

As mentioned in Chapter 1, primary batteries are not easily or safely rechargeable, and consequently are discharged and then disposed of. Many of these are "dry cells" – cells in which the electrolyte is not a liquid but a paste or similar. The electrochemical reactions that occur are not easily reversible and the cell is operated until the active component in one or both the electrodes is exhausted. Any attempt for reversing the reaction via recharging in a primary cell is dangerous and can cause the battery to explode.

The electrical resistance in primary cells is usually high; thus, even if charging was possible, it would be a slow process. At normal practical charging rates, a large proportion of the current would be dissipated as heat, causing further safety hazards. Primary batteries are therefore designed to operate at low currents and have a long lifetime. Generally primary batteries have a higher capacity (Ah/kg), a higher specific energy (Wh/kg), and a higher initial voltage than secondary (rechargeable) batteries of comparable chemistries. They are used in portable devices, toys, watches, hearing aids, and medical implants. The commercially used primary batteries are shown in Table 2.1.

2.2
The Early Batteries

Before describing in detail the main battery systems indicated in Table 2.1, the predecessors of today's batteries are introduced. Historically, the first battery was invented by Alessandro Volta, an Italian physicist, in 1800, who observed Luigi Galvani's (also an Italian physicist) experiment on frog legs in 1780 by connecting two different metals in series with the frog's leg and to one another. The leg twitched when it was touched with the iron scalpel (Figure 2.1) and, therefore, Galvani believed that frog legs can generate electricity and called it "animal electricity." However, Volta believed that the electricity came from the connection

High Energy Density Lithium Batteries. Edited by Katerina E. Aifantis, Stephen A. Hackney, and R. Vasant Kumar
© 2010 WILEY-VCH Verlag GmbH & Co. KGaA, Weinheim
ISBN: 978-3-527-32407-1

Table 2.1 Primary battery chemistries used today [1].

System cathode/ anode	Nominal cell voltage(V)	Specific energy (Wh/kg)	Advantages	Disadvantages	Applications
Carbon/ Zinc	1.50	65	Lowest cost; variety of shapes and sizes	Low energy density; poor low-temperature performance	Torches; radios; electronic toys and games
Mg/ MnO_2	1.60	105	Higher capacity than C/Zn; good shelf-life	High gassing on discharge; delayed voltage	Military and aircraft receiver-transmitters
Zn/Alk/ MnO_2	1.50	95	Higher capacity than C/Zn; good low-temperature performance	Moderate cost	Personal stereos; calculators; radio; television
Zn/HgO	1.35	105	High energy density; flat discharge; stable voltage	Expensive; energy density only moderate	Hearing aids; pacemakers; photography; military sensors/ detectors
Cd/HgO	0.90	45	Good high and low-temperature performance; good shelf-life	Expensive; low energy density	
Zn/Ag_2O	1.50	130	High energy density, good high rate performance	Expensive (but cost effective for niche applications)	Watches; photography; larger space applications
Zn/air	1.50	290	High energy density; long shelf-life	Dependent on environment; limited power output	Watches; hearing aids; railway signals; electric fences
$Li/SOCl_2$	3.60	300	High energy density; long shelf-life	Only low to moderate rate applications	Memory devices; standby electrical power devices; medical devices
Li/SO_2	3.00	280	High energy density; best low-temperature performance; long shelf-life	High-cost pressurized system	Military and special industrial needs
Li/MnO_2	3.00	200	High energy density; good low-temperature performance; cost effective	Small in size, only low-drain applications	Electrical medical devices; memory circuits; fusing

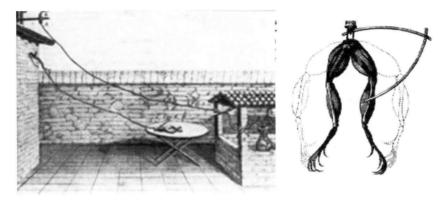

Figure 2.1 Galvani's experiment on frog legs; taken from [2].

Figure 2.2 An illustration of Volta's cell, called a "Voltaic Pile"; taken from [3].

of the two metals and proved his idea by constructing a battery consisting of alternate copper and zinc disks; a cloth or cardboard soaked in brine (NaCl) was placed in between the copper and zinc metal layers. Due to this pile-up design Volta's cell was called a "Voltaic Pile" (Figure 2.2). This cell was stable during continuous draining but electrolyte leaking was encountered. Furthermore, a thin layer of hydrogen gas (H_2) formed on the copper surface increasing the internal resistance of the battery. As a result the "Voltaic Pile" had a short life (1 h). The voltage per element of the Voltaic Pile was 0.75 V.

In 1836, John Frederic Daniel, a British chemist, improved the Volta cell by constructing a battery that used the same electrode-base metals, but a different configuration. The zinc was immersed in a sulfuric acid (H_2SO_4) solution which was withheld in a porous earthenware container. This compartment, which functioned as the anode, was then immersed in a copper sulfate ($CuSO_4$) solution that was in a copper pot that acted as the cathode. This cell, referred to as the Daniel cell, provided 1.1 V and was used for powering the telegraph system.

Eight years later in 1844, the Grove cell was invented by William Robert Grove, a British lawyer, judge, and physical scientist. The cell consisted again of a zinc anode, but the copper cathode was replaced by platinum. The anode and cathode compartments were separated by porous earthenware, and each electrode was immersed, respectively, in sulfuric and nitric acid (HNO_3) solutions.

Due to the high cost of platinum metal and the toxicity of nitric oxide (NO) gas that developed during cell operation, the grove cell was replaced by the gravity cell in the 1860s. The gravity cell was invented by Callaud, a French scientist. Callaud was able to improve the Daniel cell by excluding the porous barrier (hence reducing, the internal resistance of the system) and using a simple glass jar as the cell container. The copper cathode and zinc anode were made in a crow's foot shape and were placed at the bottom and at the top of the solution, respectively (Figure 2.3). Both electrodes were immersed in a copper sulfate solution. During cell operation, a clear solution of zinc sulfate ($ZnSO_4$) was formed around the anode and was separated from the blue copper sulfate solution by its lower density. Due to the unique shape of the electrodes this cell was also called a "crowfoot cell." The open top cell design allowed the solutions to easily mix or spill, and therefore it could be only used for stationary applications. The voltage of this battery was 1.08 V and its specific lifetime was longer than that of the voltaic pile.

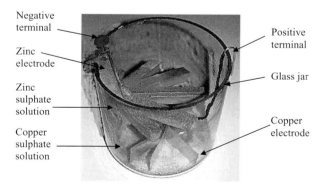

Figure 2.3 Illustration of a crowfoot cell; taken from [3].

2.3
The Zinc/Carbon Cell

Although the initial electrochemical cells consisted of zinc and copper electrodes, in the development of the first battery, the copper was replaced by carbon. The zinc-based cell is not only the first prototype battery invented but also the most widely used in our days, due to its low cost production. Its main components have not changed over the centuries, and only minor alterations were performed before reaching today's form. In the first zinc-based cell the anode was a zinc–amalgam, instead of pure zinc and the cathode contained manganese dioxide and carbon; this was the Leclanché cell.

2.3.1
The Leclanché Cell

The first primary battery was invented in 1866 by a French telegraphic engineer named Georges-Lionel Leclanché (Figure 2.4). Early telephones also used this cell as a power source. It was the first cell to contain only one low-corrosive fluid electrolyte with a solid cathode. This gave it a low self-discharge – a capacity loss without the electrodes being connected – in comparison to previously attempted batteries. The cell was called the Leclanché wet cell because an ammonium chloride (NH_4Cl) solution was used as the electrolyte. The other cell components were a zinc–amalgam (Zn/Hg) bar anode and a cathode comprising of manganese dioxide (MnO_2) and some small amount of carbon (C) powders. The cell was

Figure 2.4 Georges-Lionel Leclanché (1839–1882). (Reproduced from http://www.jergym. hiedu.cz/~canovm/objevite/objev4/leca.htm).

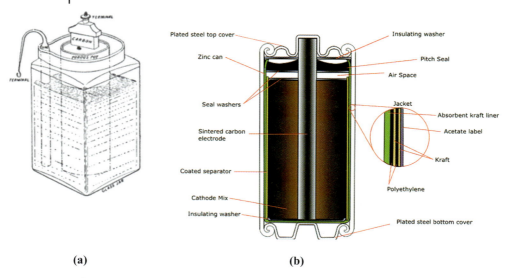

(a)   **(b)**

Figure 2.5 (a) Original Leclanché cell; taken from [4]. (b) The commercial design structure of a zinc–carbon cell; taken from [1].

designed by placing the cathode material in a porous pot and subsequently placing a carbon rod in the middle of the cathode; then the whole electrode was put in a square glass container. After inserting the zinc–amalgam bar into the same glass container the container was filled with ammonium chloride electrolyte to complete the single functional cell. An illustration of an original Leclanché cell is shown in Figure 2.5a. Leclanché improved his cell in 1876 by heating compressed manganese dioxide and carbon with resin at 100 °C to form a hard solid cathode; the porous pot was therefore not needed. The cathode was also referred to as an "agglomerate block," since a pair of agglomerate blocks were attached to the carbon plate with rubber bands and then placed in the glass container along with the zinc bar. Leclanché used his invention widely in his telegraphic office for telegraphy signaling. Comercially the most popular use of the Leclanché cell was for house door bells but later in the 1920s for portable radios, flashlights, and telephones. The voltage of this cell was 1.5 V and the capacity was 224 Ah/kg.

2.3.2
The Gassner Cell

In 1888, Carl Gassner, a German scientist, further improved the Leclanché cell by using ferric hydroxide [$FeO(OH)$] mixed with manganese dioxide and a small amount of carbon as the cathode material, and instead of using a zinc–amalgam bar, a hollowed zinc cylindrical container (bobbin) was used as the anode. The liquid ammonium chloride electrolyte was replaced by an electrolyte paste which consisted of zinc chloride ($ZnCl_2$), ammonium chloride, water, and wheat flour. The battery was therefore no longer a "wet" cell but a "dry" cell. The electrolyte paste was then mixed with the manganese dioxide/carbon mixture (cathode) and poured in the zinc anode container with the carbon rod placed in the center of the paste. The advantage of this dry cell over the Leclanché wet cell was the ease in handling and the more durable design. The successful improvement of Gassner resulted into the commercialization of this cell by the name of the zinc–carbon dry cell in 1896 by the National Carbon Company (USA). This cell provided a potential of 1.5 V, a gravimetric energy density of about 55–77 Wh/kg or 120–152 Wh/dm^3 and a capacity of 146–202 Ah/kg.

2.3.3
Current Zinc/Carbon Cell

Figure 2.5b depicts the most common commercial design of the zinc–carbon battery. The Zn serves as both the battery casing and the anode. The cathode consists of manganese dioxide (MnO$_2$), powdered black carbon, and some electrolyte (the electrolyte is ammonium chloride, while for extra heavy duty applications, the electrolyte is zinc chloride mixed with a minor amount of ammonium chloride). The MnO$_2$ and carbon mixture is wetted with electrolyte and shaped into a cylinder with a small hole in the center. A carbon rod is inserted into the center, which serves as a current collector (a material that is used to conduct electricity between the reacting parts of the electrode and the terminal and, hence, needs to be in good contact with the cathode). It is porous to allow gases to escape and provides structural support. The separator is either made from natural cellulose fibers, cereal paste coated heavy paper or from treated absorbent kraft paper (the kind of brown paper used to make large envelopes or grocery bags). The MnO$_2$ to C ratios vary between 10:1 and 3:1, with a 1:1 mixture being used for photoflash batteries, as this gives a better performance for intermittent use with high bursts of current. It should be noted that the actual cathode material is MnO$_2$, derived from natural ore; C is added for conductivity and to provide stability with the electrolyte.

The voltage of these batteries is between 1.5 and 1.7 V, their shelf-life is 1–2 years at room temperature and their service life is 110 min of continuous use (Figure 2.6 depicts a typical discharge curve). The capacity is around 40 Ah/kg at

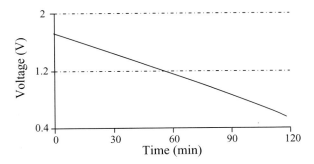

Figure 2.6 Discharge curve for a Zn/C cell at 500 mA; taken from [1].

low drains and the energy density is 55–77 Wh/kg. These cells are generally sensitive to external factors such as temperature, and are more effective when discharged intermittently. This allows for diffusion to take place so as to avoid polarization (Chapter 1, Section 1.2.2) that can occur from discharge – the separation and grouping of reactants and products which may result in increasing the internal resistance in the cell. Generally the voltage is very sloped during discharge due to cell chemistry and internal leakage. In order to minimize leakage, in modern cells, the electrolyte is produced in a gel form or held in a porous separator.

In addition to cylindrical Zn/C cells, flat Zn/C cells are also quite common. These Zn/C flat cells were introduced in the 1920s for use in portable radios. The zinc anode was painted with a carbon-conductive paint, which acted as the conductor at the anode–electrolyte interface to minimize electrical resistance. The cathode was made from a mixture of MnO_2, graphite or acetylene black carbon, and electrolyte paste. Each cell was placed in a plastic envelope and stacked inside a steel jacket (as they obeyed the basic flat cell configuration shown in Figure 1.3 of Chapter 1). Common connector strips of copper were used to collect and connect the positive and the negative electrodes to a common positive and a negative terminal. They obeyed the basic flat cell configuration shown in Figure 1.3 of Chapter 1. Zn/C flat cells are still used today for toys and some flashlights. This cell provides a potential of 1.5 V and a capacity of 238 Ah/kg.

2.3.3.1 Electrochemical Reactions

As mentioned previously the zinc/carbon cell uses a zinc anode and a manganese dioxide cathode. The carbon is added to the cathode to increase conductivity and retain moisture; therefore, only manganese dioxide, zinc, and electrolyte take part in the reaction.

The overall reaction in the cell is

$$Zn + 2MnO_2 \rightarrow ZnO \cdot Mn_2O_3 \quad +1.54 \text{ V}$$

The exact mechanism for this is complicated, and not fully understood; however, the approximate half-cell reactions are

$$\text{Anode:} Zn \rightarrow Zn^{2+} + 2e^- \qquad\qquad +0.76 \text{ V}$$

$$\text{Cathode:} 2NH_4^+ + 2MnO_2 + 2e^- \rightarrow Mn_2O_3 + H_2O + 2NH_3 \quad +0.78 \text{ V}$$

The electrolyte in which the above reactions take place is ammonium chloride. However, this is complicated by the fact that the ammonium ion produces two gaseous products:

$$2NH_4^+ + 2e^- \rightarrow 2NH_3 + H_2$$

These products must be absorbed in order to prevent build up of pressure in the vessel, therefore zinc chloride is added to the ammonium chloride in order for the following two reactions to occur:

$$ZnCl_2 + 2NH_3 \rightarrow Zn(NH_3)_2 Cl_2$$

$$2MnO_2 + H_2 \rightarrow Mn_2O_3 + H_2O$$

2.3.3.2 **Components**

Historically the black carbon used was in the form of graphite, however acetylene black carbon is often used in modern batteries as it can hold more electrolyte. Various forms of MnO_2 are available: (i) natural manganese dioxide; ores occur naturally in Gabon, Greece, and Mexico with 70–85% MnO_2, (ii) activated manganese dioxide, (iii) chemically synthetic manganese dioxide (90–95% MnO_2), (iv) electrolytic manganese dioxide (EMD) is widely used in industrial cells but mainly in alkaline batteries as it allows for a higher cell capacity, higher rate capabilities, and less polarization.

Pure Zn sometimes alloyed with minor alloying elements such as Pb is used as the can material thus serving both as the anode and as the container. The electrolyte for a standard zinc/carbon (Leclanche) cell uses a mixture of ammonium chloride and zinc chloride in an aqueous solution. A zinc-corrosion inhibitor is also added, which forms an oxide layer. This inhibitor is usually mercuric oxide or mercurous chloride. A typical electrolyte composition is 26.0% NH_4Cl, 8.8% $ZnCl_2$, 65.2% H_2O, 0.25–1.0% corrosion inhibitor. While the oxide layer prevents corrosion, thus prolonging the shelf-life, it increases the electrical resistance permitting use only under low currents.

The carbon rod is inserted into the cathode and acts as a current collector. It also provides structural support and vents hydrogen gas that evolves as the reactions proceed. The as produced rods are very porous, so they must be treated with waxes or oils to prevent loss of water, but at the same time remain porous enough to allow hydrogen to pass through. Ideally they should also prevent oxygen entering the cell, as this would aid corrosion of the zinc.

The separator physically separates the anode and the cathode, but allows ionic conduction to occur in the electrolyte. The following two types of separators are used: (i) a gelled paste which is placed into the zinc can, and once the cathode is inserted, the paste is forced up the sides of the can between the zinc and the MnO_2 cathode; (ii) kraft paper coated with cereal paste, or another gelling agent, rolled into a cylinder and along with a circular bottom sheet is added to the can. The cathode is added, and the rod inserted, pushing the paper against the walls of the cans. This compression releases some electrolyte from the cathode mix, soaking the paper. As the paste is relatively thick, more electrolyte can be held by the paper than the paste, giving an increased capacity; thus paper is usually the preferred separator.

The seal of the battery can be comprised of asphalt pitch, wax/resin mix, or plastic (usually polyethylene or polypropylene). Air space is usually left between the seal and the cathode to allow for expansion. The function of the seal is to prevent evaporation of the electrolyte, and to prevent oxygen entering the cell and corroding the zinc.

The jacket provides strength and protection, and will hold the manufacturers' label. It contains various components, which can be metal, paper, plastic, mylar (biaxially oriented polyethylene terephthalate [boPET] polyester film), cardboard (sometimes asphalt-lined), and foil. It should be noted here that the jacket is not an active material in the cell but serves as the cathode current collector.

The terminals of the battery are tin plated steel or brass. They aid conductivity and prevent exposure of the zinc.

2.3.4
Disadvantages

There are various shapes and sizes of zinc–carbon dry cells, but the most common ones fabricated are of size D (see Table 1.1 of Chapter 1). They have been used in flashlights (familiar name "flashlight cell"), portable radios, photoflashes, and toys since 1896 and are still manufactured today (Energizer Company). The reason for producing large sizes is the physical limitation by which cell efficiency increases by decreasing the current density; this corresponds to a large cell size [5]. Disadvantages in zinc–carbon batteries arise from the fact that electrolyte appears in the overall cell reaction, resulting in a decreasing electrolyte concentration during cell operation, and causing corrosion of the zinc anode by acidic NH_4^+ ions [6]. As a result such batteries have an unstable voltage and short battery shelf-life.

2.4
Alkaline Batteries

Research on zinc–carbon dry cells has improved their shelf-life and power density for heavy duty devices such as portable lighting, motors, and has met the environmental benign requirements. The improvements that were performed on zinc–carbon batteries in order to make them more efficient resulted in the development of alkaline-manganese dioxide batteries, which are the most common commercial dry cells.

Lewis Urry developed the small alkaline battery at the Ever Ready company laboratory in 1949 and this cell design was further improved by Samuel Ruben in 1953. The first modern alkaline cell was commercialized in 1959 by Ever Ready, and by 1970 it was produced all over the world. Currently over 15 billion alkaline cells are used worldwide each year. They use the same anode and cathode as the zinc/carbon cells, but their electrolyte contains an alkaline solution (aqueous KOH), and, hence, their name alkaline cells. Their capacity at over 65 Ah/kg is 25–30% higher than the zinc/carbon cell. Furthermore, they have a very good performance at high discharge rates and a continuous discharge at low temperatures.

The active materials used are essentially the same as in the Gassner cell: zinc and manganese dioxide and C. However, the electrolyte is potassium hydroxide, which is very conductive, resulting in a low internal electrical resistance for the cell due to the higher ionic conductivity. In particular, synthetic manganese dioxide (electrolytic manganese dioxide (EMD) or chemically synthetic manganese dioxide (CMD)) replaced natural manganese dioxide ore because of its higher capacity and activity over natural material, while black carbon replaced graphite in the anode since it has superior conducting and absorption properties [7]. Furthermore, in order to increase the surface area of the anode, the cell design utilizing a zinc can

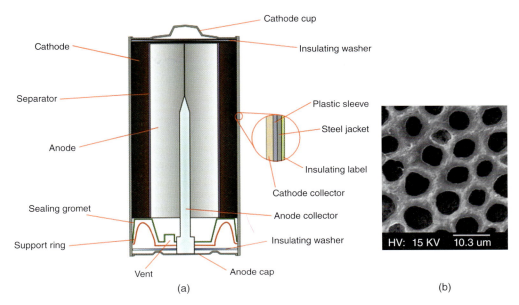

Figure 2.7 (a) The commercial design structure of an alkaline cell, (b) microporous separator, taken from [1].

(which serves as a conductor and a mechanically strong container) as the anode was replaced with a cell design using an external steel jacket and a gel of amalgamated zinc (Zn/Hg) powder mixed with electrolyte as the anode. This resulted in a better ionic conduction and reduced the mass-transport polarization (Figure 2.7). The current collector for this anode is a brass (CuZn) pin that is inserted through the middle of the gel and connected to the anode terminal. The steel can is nickel-plated or has a conductive carbon coating placed on the inner surface (Figure 2.7a).

It can be seen by comparing Figures 2.5b and 2.7a that alkaline cells are "inside out" compared to the Zn/C cells since the manganese dioxide cathode is external to the zinc anode, giving better diffusion properties, and lower internal resistance. The container is made of steel which is readily passivated in an alkaline solution and serves as both a conductor and a mechanically strong container. The entire cell is hermetically sealed with nylon.

Additional advantages of alkaline batteries are: they have a longer service delivery over zinc–carbon cells; their capacity is constant during varying discharge loads; they can be heavily drained for long periods of time. The operation temperature is in the range of −20 to +70°C, the gravimetric energy density is 66–99 Wh/kg or 122–268 Wh/dm^3 and the total capacity can range from 700 mAh to 10 Ah for multiple-cell batteries with different voltages; multi-cell batteries are made by connecting miniature or cylindrical cells in series or parallel. A typical

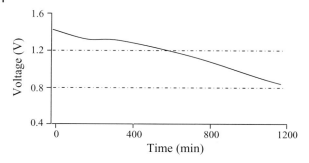

Figure 2.8 Discharge curve at 500 mA for an alkaline cell; taken from [1].

discharge curve is shown in Figure 2.8. The typical commercial service capacity is higher than the zinc–carbon battery (whose gravimetric energy density as mentioned before is about 55–77 Wh/kg or 120–152 Wh/dm^3). The commercial sizes of alkaline batteries vary from N, AA, AAA, C, and D. Small portable electronic devices such as MP3 players, digital cameras, and scientific calculators are examples of devices employing general AA or AAA-size alkaline batteries. Alkaline cells are not manufactured in a flat cell configuration, but they do come as button cells.

2.4.1
Electrochemical Reactions

The following half-cell reactions take place inside the cell:

At the anode: $Zn + 2OH^- \rightarrow Zn(OH)_2 + 2e^-$ +1.25 V

$Zn(OH)_2 + 2OH^- \rightarrow [Zn(OH)_4]^{2-}$

At the cathode: $2MnO_2 + H_2O + 2e^- \rightarrow Mn_2O_3 + 2OH^-$ +0.4 V

For full discharge: $MnO_2 + 2H_2O + 2e^- \rightarrow Mn(OH)_2 + 2OH^-$

Overall: $Zn + 2MnO_2 \rightarrow ZnO + Mn_2O_3$

or $Zn + MnO_2 + 2H_2O \rightarrow Mn(OH)_2 + Zn(OH)_2.$ +1.65 V

It is not possible to describe the cathodic reaction on discharge in a simple unambiguous way, despite a lot of research. In fact the discharge curve has two fairly distinct sections corresponding to the change in the oxidation state of Mn from +4 to +3 and then to +2 during the reduction of MnO_2. The reality is more complicated than described in the two reactions shown above.

2.4.2
Components

For an alkaline cell, electrochemically produced MnO_2 must be used. The ore rhodochrosite ($MnCO_3$) is dissolved in sulfuric acid, and electrolysis is carried out under carefully controlled conditions using titanium, lead alloys, or carbon for the electrode onto which the oxide is deposited. This gives the highest possible purity, typically $92 \pm 0.3\%$. The cathode itself also contains approximately 10% graphite – a higher %C results in more powerful batteries. A typical composition of a cathode would be 70% MnO_2 (of which 10% is water), ~10% graphite, 1–2% acetylene black carbon, with some binding agents and electrolyte for aiding wetting with the electrolyte by decreasing the surface tension.

The zinc anode must be very pure (99.85–99.90%) and is produced by electroplating or distilling. A small amount of lead is sometimes added to help prevent corrosion (usually ~0.05%). The zinc is powdered by discharging a small stream of molten zinc into a jet of air "atomizing" it. The powder contains particles between 0.0075 and 0.8 mm. In particular, there are two methods for the formation of the anodes from the powder:

- **Gelled anodes:** These contain approximately 76% zinc, 7% mercury, 6% sodium carboxymethyl cellulose, and 11% KOH solution, which is extruded into the cell as the viscosity is high. In very small cells NaOH is added to reduce creepage around the seal area. However, this mixture is not ideal as it does not fully utilize the zinc at high current densities. Two phase anodes have therefore been developed, consisting of a clear gel phase and a more compact zinc-powder gel phase, which enables 90% zinc usage.

- **Porous anodes:** The zinc powder is wetted with mercury and cold pressed, welding the particles together. The porosity can be controlled by materials such as NH_4Cl or plastic binders if required, which can be removed later. These anodes can carry very high currents.

The separators are "microporous" and are made from woven or felted materials. When nonwoven separators made from synthetic fibers are used, they must be selected to resist high pH values of the alkaline electrolyte. An example of a polymeric membrane separator is shown in Figure 2.7b.

2.4.3
Disadvantages

Zinc can progressively corrode in the alkaline electrolyte, releasing hydrogen gas, which results in a pressure increase and the consequent expansion of the cell. Such a reaction can be suppressed by alloying Zn with elements such as Hg or Pb; however, both are notorious for their toxicity. While the alkaline cells are superior to the Zn–C dry cells, for their steadier voltage output, their longer life, and their higher current capability, they still have a fairly high internal resistance. Their components also have a high thermal coefficient of resistivity. As a result of these factors, the faster an alkaline cell drained, the higher the percentage of the load dissipated as heat. The capacity is therefore greatly reduced. For example, with a AA-sized alkaline battery, the nominal capacity is >3000 mAh/kg at low drains (<500 mA), but at a current drain of >1000 mA (as may be typically required in a digital camera), the capacity will be quite low at <750 mAh/kg.

2.5
Button Batteries

In addition to cylindrical batteries there is another category of batteries that are known as button batteries. The common feature between material chemistries used in button batteries is that they allow for low discharge rates, and therefore provide a longer lifetime, making them suitable for scientific and biomedical implantable devices. Their common design is shown in Figure 2.9, while descriptions of button cells are given below.

2.5.1
Mercury Oxide Battery

The first button cell was the mercuric oxide (HgO) battery, which was developed during World War II for military applications and was then used in scientific instruments. This cell comprises of a zinc anode, and a mercuric oxide cathode, while the electrolyte consists of zinc oxide mixed with potassium hydroxide/sodium hydroxide (NaOH) (the zinc oxide makes up 7% of the weight of the potassium hydroxide electrolyte). Because of its high performance, such as high capacity per volume (300–500 W/dm³), stable voltage upon discharge, and long shelf-life, this cell was used in hearing aids, watches, cameras, and scientific calculators. There are two battery designs: a 1.35 V, whose cathode is 100% mercuric oxide, and a 1.4 V, whose cathode is a mixture of mercuric oxide and manganese dioxide.

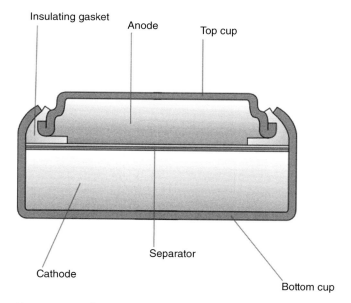

Figure 2.9 Configuration of a button battery; taken from [1].

The first cell is suitable for stable voltage and long storage applications such as sonobuoys (underwater sonic wave receivers) and early satellites, while the second one is most suitable for long-term continuous low-drain applications, such as hearing aids and watches. Due to the environmental concern on using mercury metal, this battery was replaced by zinc/silver, zinc–air, and lithium batteries. In addition to button cells, there also exist flat-pellet and cylindrical configurations, but they are expensive and used only for special applications.

Electrochemical Reactions
The reactions that take place upon discharge of these cells are as follows
  At the anode:

$$Zn + 2OH^- \rightarrow ZnO + H_2O + 2e^- \quad +1.25 \text{ V}$$

at the cathode:

$$HgO + H_2O + 2e^- \rightarrow Hg + 2(OH)^- \quad +0.0977 \text{ V}$$

overall:

$$Zn + HgO \rightarrow ZnO + Hg \quad +1.35 \text{ V.}$$

A typical discharge curve is shown in Figure 2.10.

2.5.2
Zn/Ag$_2$O Battery

The first zinc/silver oxide battery was developed in the late 1930s by G. André by using a suitable separator that could clearly separate the two electrodes. These cells have a relatively high specific energy density of 150 Wh/kg, and can deliver current at a very high rate, with constant voltage of 1.6 V. However, the construction materials are of high cost, so such battery systems are limited to the production of button cells, for use in calculators, watches, hearing aids, and other such devices that require small batteries and a long service life. A typical discharge curve is shown in Figure 2.11.

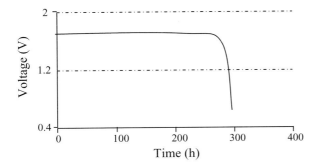

Figure 2.10 Discharge curve at 6.5 kΩ for a HgO cell; taken from [1].

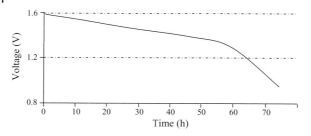

Figure 2.11 Discharge curve at 1 kΩ for a Zn/Ag cell; taken from [1].

The cathode is generally composed of monovalent silver oxide with added graphite to improve conductivity. The anode is zinc powder mixed with a gelling agent, which is then dissolved in the alkaline electrolyte; these batteries however are not referred to as alkaline cells, although scientifically they are also a type of alkaline cell. The two electrodes are separated by a combination of layers of grafted plastic membrane, treated cellophane, and nonwoven absorbent fibers. The top cup (negative terminal) is made up of laminated layers of copper, tin, steel, and nickel, and the bottom cup (positive terminal) is nickel-plated steel. An insulating gasket prevents contact between the two. The cell looks similar to the HgO button cell as shown in Figure 2.9.

Electrochemical Reactions
The reactions that occur in this cell are indicated below.
At the anode:

$$Zn + 2OH^- \rightarrow Zn(OH)_2 + 2e^- \quad +1.25 \text{ V}$$

at the cathode:

$$Ag_2O + H_2O + 2e^- \rightarrow 2Ag + 2OH^- \quad +0.35 \text{ V}$$

overall:

$$Ag_2O + H_2O + Zn \rightarrow 2Ag + Zn(OH)_2 \quad +1.6 \text{ V}$$

The voltage of these batteries is around 1.6 V. It is linearly dependent on temperature, while they can be used continuously for several thousand of hours and their shelf-life can be several years at room temperature. Their discharge characteristics are relatively flat with a specific capacity of 80–90 Ah/kg.

2.5.3
Metal–Air Batteries

When a reactive metal anode is electrochemically coupled with an air (oxygen) cathode, the cathode provides an exhaustible amount of nearly weightless reactive component. A basic metal–air battery consists of a metal anode, an oxygen cathode, and an electrolyte through which ions can diffuse and electrochemically oxidize the anode. While oxygen is consumed at the cathode, it is quickly replenished from the surrounding air; thus, there is no material consumption at the cathode. Oxygen diffuses through the porous cathode and is electrochemically reduced in the cell:

$$O_2 + 2H_2O + 4e^- \rightarrow 4OH^- \quad +0.401 \text{ V}$$

The hydroxide ions then diffuse through the aqueous electrolyte (usually neutral or alkaline solutions) and oxidize the metallic anode. Such cells have a high value of theoretical energy density and provide a flat discharge voltage curve. The capacity limit is determined by the anodic capacity and the presence of any parasitic reactions. Strictly speaking, these cells are primary cells as they cannot be electrochemically recharged once the anode is exhausted. However, the spent anode can be replaced mechanically with a new fresh anode; thus, the term mechanically rechargeable battery is often used in the literature. Since air is used as an oxidant, and a metal anode is considered as a solid fuel, the term metal fuel cell is also sometimes used to describe these cells.

Considerable progress has been made in the development of air electrodes. A metallic mesh is used as the current collector and as a support for the cathode. An air access hole allows access to oxygen from the surrounding air. The rate of oxygen transfer into the cathode is controlled by the area of the hole and/or the porosity of a diffusion membrane deposited on the cathode surface. The operating current achieved in the cell is directly proportional to the oxygen consumption at the cathode. Thus, the current can be increased by increasing the porosity until the reaction rate of the above reaction becomes rate limiting. The catalyst layer itself consists of manganese oxide in a carbon matrix (a small platinum loading may also be provided), and is made hydrophobic by admixing with fine Teflon particles. Thus, a dry boundary is maintained between the gas phase and the aqueous electrolyte.

A number of metallic electrodes have been developed for metal–air cells. Table 2.2 provides values for theoretical cell voltages, theoretical specific capacities, and theoretical specific energies for selected metal anodes. The practical operating voltages are often much lower due to polarization effects at the electrodes.

In the subsequent subsections, two different metal/air batteries are described.

Table 2.2 Metal–air cells.

Metal anode	Theoretical capacity (Ah/g)	Theoretical cell voltage (V)	Theoretical specific energy (Wh/g)	Operating voltage (V)
Li	3.86	3.4	13	2.4
Al	2.98	2.7	8.1	1.6
Mg	2.20	3.1	6.8	1.4
Ca	1.34	3.4	4.6	2.0
Zn	0.82	1.6	1.3	1.2
Fe	0.96	1.3	1.2	1.0

2.5.3.1 **Zn/Air Battery**

The zinc/air system has attracted much attention as a candidate for electrical vehicle (EV) applications. It is also seen as an environmentally friendly cell. The development of this battery began in the 1930s and was used for military radios in World War II. It was finalized in today's commercial design in the early 1970s in N and D sizes, which are mostly used for paging equipment. Later on, the button cell was introduced for hearing aids and watches. The chemistry design of this cell was adopted from the research on zinc–air fuel cells, which comprised of a zinc granular powder and electrolyte mixture anode, a hydrophobic cathode composite (low surface black carbon and Teflon (polytetrafluoroethylene, C_nF_{2n} (PTFE)) particles), a catalyst-coated nickel (Ni) screen layer, an air diffusion Teflon layer, a hydrophobic Teflon layer, and an air distribution membrane. This cell has air access holes allowing oxygen from the ambient atmosphere to enter the cell and perform the electrochemical reaction at the cathode. These batteries are commercially available in 1.4 V single-cell, and 5.6 V and 8.4 V multicell batteries. The energy density of the 1.4 V batteries ranges from 442 to 970 Wh/dm^3. These batteries, therefore, are not only preferable from an environmental point of view over zinc/silver, but are also more efficient. In addition to the button configuration, Zn/air batteries are also massively produced in the flat cell configuration.

Electrochemical Reactions The electrochemical reactions that take place in this cell are as follows:
At the anode:

$$Zn \rightarrow Zn^{2+} + 2e^- \quad +1.25 \text{ V}$$

$$Zn^{2+} + 2OH^- \rightarrow Zn(OH)_2$$

$$Zn(OH)_2 \rightarrow ZnO + H_2O$$

at the cathode:

$$\tfrac{1}{2}O_2 + H_2O + 2e^- \rightarrow 2OH^- \quad +0.40 \text{ V}$$

overall:

$$Zn + \tfrac{1}{2}O_2 \rightarrow ZnO \quad +1.65 \text{ V}.$$

In addition to providing high energy densities, the zinc/air cells have a flat discharge voltage, after the initial fall from 1.4 to 1.2 V (Figure 2.12) and a long shelf-life. The button cell configurations have been successfully applied in hearing aids, watches, and other low-power applications such as memory preservation. For electric vehicle applications, only modest progress has been made due to several operational difficulties.

Disadvantages The service life of a Zn/air battery is affected by the "water vapor transfer." It arises from the inevitable vapor pressure difference between the atmosphere and over the electrolyte. At low humidity, the electrolyte can get con-

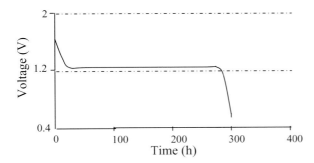

Figure 2.12 Discharge curve at 620 Ω for a Zn/air cell; taken from [1].

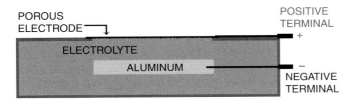

Figure 2.13 Schematic depiction of a standard aluminum-air battery; taken from [1].

centrated causing cell failure due to electrolyte loss, while at high humidity the electrolyte is diluted resulting in a lower conductivity. Under sustained water gain, the cathode catalyst is easily flooded accompanied by a loss of electrochemical activity. Direct oxidation of the zinc anode by oxygen entering the cell will result in a loss of capacity, accentuated at higher temperatures. The alkaline solution is also prone to carbonation due to direct reaction between the alkaline electrolyte and the atmospheric carbon dioxide (380 ppm), which can aggravate a water loss reaction and impede the air access at the cathode.

2.5.3.2 Aluminum/Air Batteries

The structure of a standard aluminum/air battery is illustrated in Figure 2.13. A piece of aluminum is immersed in an electrolyte near a porous electrode. This porous electrode has air on one side and the electrolyte on the other. The electrolyte can be an aqueous NaCl solution (or sea water) or an alkali solution such as potassium hydroxide.

The aluminum reacts with OH^- ions to form aluminum hydroxide and release three electrons. The OH^- ions are present either because the electrolyte is an alkali solution, or because they are produced at the other electrode (see below). In particular, the reaction is

$$Al + 3OH^- \rightarrow Al(OH)_3 + 3e^- \quad +2.30 \text{ V}$$

At a very high pH, the $Al(OH)_3$ may dissolve in the electrolyte and the corresponding electrochemical reaction may be expressed as

$$Al + 4OH^- \rightarrow Al(OH)_4^- + 3e^- \quad +2.31 \text{ V}$$

At the porous electrode (cathode), which is often made of carbon, the water in the electrolyte reacts with oxygen from the air and absorbs the electrons produced at the aluminum electrode as

$$O_2 + 2H_2O + 4e^- \rightarrow 4OH^- \quad +0.401 \text{ V}.$$

With a salt water electrolyte, the open-circuit voltage of the cell is only about 1.2 V, but the normal operating voltage is much lower at about 0.7 V or 0.8 V. With the KOH solution, the voltages are about 0.5 V higher and further optimized air electrodes for use with alkali solution are commercially available.

The energy density of the aluminum/air battery is excellent, but it is not yet in extensive use. The main reason for this is that corrosion reactions can take place between the electrolyte and the aluminum, leading to hydrogen evolution. If this happens at a high rate during operation, then clearly the main power-generating function of the battery will be impaired, but even a low rate of reaction will cause damage to the anode over time and this can happen during storage. However, if the electrolyte is stored separately, and added when the power is needed, then this problem can be avoided. This type of battery is usually called a reserve battery, and this is one market in which aluminum/air batteries have had some success. It has been reported that corrosion of aluminum can be minimized by using special alloying elements and/or by using special additives in the electrolyte.

2.6
Li Primary Batteries

The demand for high energy density power sources to be used in small size and light weight equipment, led researchers to explore Li-based cells since Li is the lightest metal. The first Li battery system was introduced in 1958 by W. S. Harris, who studied the solubility and conductivity of inorganic salts in the cyclic esters, ethylene carbonate, propylene carbonate, gamma-butyrolactone, and gamma-valerolactone. Li is the anode, while some form of C (as the current conductor) is mixed with the electroactive cathode component such as manganese dioxide or thionyl chloride, etc [8].

Apart from the fact that Li is the lightest of all metals and it can float on water, it is also the most electropositive metal with a very low standard reduction potential of −3.05 V at 298 K; thus, a very high capacity of 3860 Ah/kg can be obtained from Li. Other battery-base metals give off much lower capacities; sodium provides 1.16 Ah/g, while zinc gives 0.82 Ah/g. The drawback with Li, however, is that it can react very violently with water and air, prohibiting, thus, the use of aqueous electrolytes under normal conditions. Due to the reactivity of Li, all Li battery systems pay particular attention to abuse-free sealing (i.e., the consumer cannot easily tamper with the sealing by accident or on purpose) and rely mainly on organic electrolytes which do not have as high a conductivity as aqueous electrolytes. Lower conductivity is advantageous since it allows for a low self-discharge and, therefore, long life batteries can be obtained, which are particularly useful in biomedical implantable devices. With many electrolytes such as propylene carbonate, liquid sulfur dioxide, thionyl chloride, and some molten salts, Li forms a passivating layer called the solid electrolyte interface, which prevents corrosion but not cell discharge. Finding suitable mixtures of organic liquids and metal salts that were safe to use took many years. This research resulted in the production of many different kinds of cells with a wide range of applications, from watches with millimeter batteries to batteries large enough to power small cars. Cathode materials are varied but a special *graphite* with fluorine treatment and manganese dioxide are among the most common used commercially for nonrechargeable cells.

The first Li primary battery was commercially available in the form of a cylindrical lithium/manganese dioxide cell by Sanyo Electric Co. in 1975. It was used in cameras and similar devices. Another variation uses iron sulfide as the cathode. This powerful combination provides a 1.5 V battery which can be directly substituted for alkaline or Leclanche (carbon–zinc) cells.

Several lithium systems have been developed over the past three decades. The anode always comprises lithium metal while alternate materials have been employed as cathodes. Examples of lithium primary battery systems are lithium–sulfur dioxide (SO_2), lithium–manganese dioxide (MnO_2), lithium–copper fluoride (CuF_2), lithium–silver chromate (Ag_2CrO_4), lithium–lead bismuthate ($Bi_2O_3 \cdot 2PbO$), lithium–polycarbon monofluoride ($(CF)_x$), lithium–iodine (I). However, lithium primary batteries cannot compete with alkaline batteries in the market, due to the fact that they provide the same life as alkaline batteries, but at much higher manufacturing costs, and at the expense of safety issues. In particular, the manufacturing of lithium cells and batteries is a high technology business. The cells must be assembled in rooms with relative humidity below 3% and preferably equal to or less than 1%. Many major battery manufacturers have had serious factory fires when the materials of the batteries were unintentionally mishandled. Two Li battery systems are described below.

2.6.1
Lithium/Thionyl Chloride Batteries

In early 1968, a Li-primary cell was developed that allowed for a higher energy density than the previous Li cells and had a service life of 15–20 years [5]. It consists of a lithium foil anode and a thionyl chloride ($SOCl_2$) cathode; both electrodes are placed in an organic electrolyte that contains lithium ion electrolyte (lithium tetrachloroaluminate, $Li(AlCl_4)$). These cells provide a very high capacity of 450 Ah/kg, and an energy density of 330 Wh/kg.

The thionyl chloride is blended with a specially chosen carbon, such that it can be used as the cathode in a lithium cell without a separator; a more flexible feature in comparison with an aqueous battery. This allows a very low internal cell resistance in a very high energy package. The aforementioned solid electrolyte insulating film reforms naturally the instant the discharge of the cell ceases. Such lithium batteries can produce a very high current for fairly long periods of time. The cells are expensive and must be assembled very carefully to avoid accidents, as they can get very hot or even catch fire if they are not assembled and handled properly. The discharge reaction for the Li/thionyl battery can be described as

$$2Li \rightarrow 2Li^+ + 2e^- \qquad\qquad\qquad\qquad 0\ V$$

$$SOCl_2 + 2Li^+ + 2e^- \rightarrow 2LiCl + \frac{1}{2}S + \frac{1}{2}SO_2 \quad +3.6\ V.$$

The overall cell reaction is

$$2Li + SOCl_2 \rightarrow 2LiCl + \frac{1}{2}S + \frac{1}{2}SO_2 \quad +3.6\ V,$$

where the thionyl chloride acts not only as the cathode material but also as the electrolyte when combined with a Li salt such as $LiClO_4$ or $LiAlCl_4$. A carbon black electrode is used as the conductor at the cathode. The cell provides a flat discharge curve for a long period of time and is therefore useful for computer backup applications. This cell has a high specific energy density, a long shelf-life with low-to-moderate drains and is well suited for medical devices such as pacemakers for low current drains, drug pumps, and neuro-stimulators with medium power requirements. A flat discharge curve with a rapid loss of voltage at the end of the discharge can be a serious problem in medical devices. A typical discharge curve is shown in Figure 2.14.

2.6.2
Lithium/Sulfur Dioxide Cells

The lithium/sulfur dioxide cell was developed in the 1960s by Honeywell Inc. and the Mallory Battery Company [5]. The electrolyte is highly conductive and comprises of acetronitrile with dissolved lithium bromide (LiBr) and sulfur dioxide (SO_2). The SO_2 participates in the cathodic reaction on a porous carbon current collector as

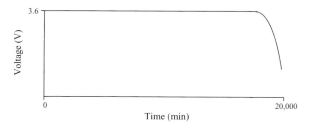

Figure 2.14 Discharge curve at $36\,k\Omega$ for a litium/thionyl cell; taken from [1].

$$2SO_2(l) + 2e^- \rightarrow S_2O_4^{2-}(l) \quad 3.0 \text{ V.}$$

High-power cells are designed to deliver pulse loads of over 30 A at a voltage of 2 V. Practical energy densities of greater than 300 Wh/kg with capacities greater than 400 Ah/kg are easily available. Overheating and thermal runaway can be a problem in such a high current/high capacity cell; thus, use of shut down separators and other external safety devices, such as fuses or current limiters, is important.

2.7
Oxyride Batteries

The most recent primary battery, launched in 2004, is the "Oxyride battery" by the Panasonic Corporation in Japan. This cell is suitable for high-drain devices such as digital cameras and cars [9]. In the same year a 2.9 m special "oxyride car" with a 50 kg passenger was powered by only two oxyride batteries. The difference of this cell from alkaline cells is that a "Tablet Mix Control System" was employed to increase the amount of cathode materials, which are an oxy nickel hydroxide (NiOOH) solution mixed with a finer grained manganese dioxide and graphite for better cell performance. A vacuum-pouring technique is applied for packing a high amount of electrolyte into the cell to increase the surface area of the electrode. Higher power, longer battery life, and higher battery durability are therefore obtained. The operating voltage of this cell is in the range of 1.5 to 1.7 V, while the capacity of the cell is 28 Ah/g for AA sizes and 25 Ah/g for AAA sizes (up to two times longer than traditional alkaline batteries). The schematic representation of the cathode and anode characteristics of alkaline compared to oxyride batteries is shown in Figure 2.15. The active element indicated in the figure is finer grained graphite powder.

As the oxyride batteries are under patent the discharge curves and half reactions that occur are disclosed.

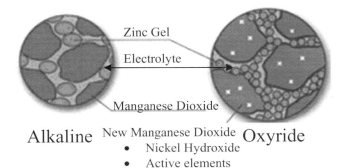

Figure 2.15 The schematic presentation of cathode and anode characteristics of alkaline and oxyride batteries; taken from [10].

2.8
Damage in Primary Batteries

Ideally the lifetime of batteries is dictated by their discharge curves, and they can be used efficiently until the galvanic cell has been drained completely. Over the lifetime of the cell, however, the morphology of the components will change; these morphological changes affect detrimentally the electrochemical properties of the cell. In particular, the following phenomena can be distinguished:

- During operation of the battery the temperature increases. When crystalline metals are exposed to elevated temperatures, their grains grow larger which results in an increase in the impedance of the cell, due to loss of contact, thus decreasing the capacity and energy density.

- The diffusion of leaking electrons through the electrolyte to the anode surface causes dendrite formation on the anode, which as a result increases the anode volume and forces the two electrodes closer together. The self-discharge of the battery therefore increases resulting in a shorter lifetime.

- A metal (anode) may dissolve in the electrolyte and then reform at the cathode as dendrites, as shown for zinc in an alkaline cell in Figure 2.16. Similar images of dendrites are shown in Figure 2.17 [11]. The scanning electron microscopy and transition electron microscopy images of the InSb anode material for lithium-ion batteries were prepared by the solution route method. The dendrites were nucleated and grown on zinc particles, which served as a reducing agent. The electron transfer process in growing the InSb, therefore, is analogous to the electron transfer process giving rise to the dendrites demonstrated in Figure 2.16.

dendrite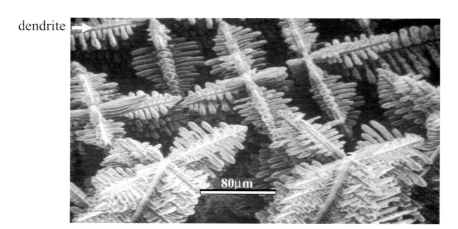

Figure 2.16   Dendrite formation on Zn anode; taken from [1].

 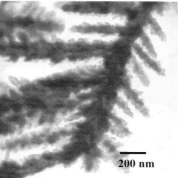

200 nm

Figure 2.17 SEM and TEM images of InSb dendrite; the InSb anode material was prepared by the solution route method for a lithium-ion battery [11].

- When dendrite formation is severe, dendrites may penetrate the separator giving a very high self-discharge, or even a short circuit of direct electron flow between the anode and the cathode bypassing the electrolyte. Short circuit in a battery is highly undesirable, as this can produce large local transient currents leading to overheating and the unleashing of destructive power in the battery. Accidental short-circuiting of such a cell can cause extremely high currents and heating that can lead to excessive temperatures at which runaway reactions can occur between the Li anode and the cathode and the components of the cell. Breakdown of chemical compounds can result in a rapid build-up of pressure within the cell resulting in explosions, fires, and serious secondary damages.

Internal short circuits do not only develop due to dendritic growth on the electrodes, but can also occur by damaged separators, as follows: In the event of overheating, the separators which are often made of polymers may be distorted or may melt (or both phenomena can take place) resulting in a short circuiting and increasing the rate of the runaway reaction; as more heat is produced, the reactions occur faster and faster. To avoid fires that may take place in such cases two precautions are taken: (i) external battery circuits are available that can cut off the device on detecting a sudden drop in voltage, (ii) chemical inhibitors are added to electrolytes to make them flame retarding or self-extinguishing by acting as sinks for heat evolved in case runaway reactions occur. In order to prevent such deficiencies from occurring, suitable separators have to be designed during manufacturing. For example, separators are made rigid so as not to readily distort or melt on heating or whose impedance increases rapidly on heating resulting in the shut down of the battery. It is also common to provide for vents to avoid a dangerous buildup of pressure inside the cell due to gassing. In general, batteries contain reactive components and have the potential to be dangerous if they are not carefully designed or handled.

2.9
Conclusions

Primary batteries are normally used in low drain applications. They are not suitable for high drain operations due to decreased lifetime and the need to be replaced. Primary batteries reproduce typically only 2% of the energy used during the manufacturing process and therefore are very inefficient from the point of energy usage and sustainability but their easy to use nature makes them attractive in many applications. In the next chapter more evolved electrochemical cells that can be recharged will be presented.

References

1 University of Cambridge (2005) DoITPoMS Teaching and Learning Packages, http://www.doitpoms.ac.uk/tlplib/batteries/index.php (accessed 5 February 2010).
2 Pierre R. Roberge (1999) http://www.corrosion-doctors.org/Biographies/GalvaniBio.htm (accessed 9 July 2009).
3 W1TP Telegraph and Scientific Instrument Museums (2009) *The Cyclopedia of Telephony and Telegraphy* (1919), Tom Perera.
4 Linden, D. and Reddy, T.B. (2002) *Handbook of Batteries*, 3rd edn, McGraw-Hill, New York.
5 Crompton, T.R. (2000) *Battery Reference Book*, 3rd edn, Newnes, Oxford.
6 Silberberg, M.S. (2000) *Chemistry: The Molecular Nature of Matter and Change*, 2nd edn, McGraw-Hill, New York

7 Pistoia, G. (2005) *Batteries for Portable Devices*, Elsevier Science B.V., Amsterdam.
8 van Schalkwijk, W.A. and Scrosati, B. (2002) *Advances in Lithium-Ion Batteries*, Klumwer Academic Publishers, Dordrecht
9 Wise GEEK homepage (2003–2010) http://www.wisegeek.com/what-is-an-oxyride-battery.htm (accessed 4 November 2007).
10 Australia Panasonic homepage (2006) panasonic.com.au (accessed 4 November 2007).
11 Sarakonsri, T., Choksawatpinyo, K., Seraphin, S., and Tunkasiri, T. (2005) *CMU J.* Special Issue on Nanotechnology, **4**, 73.

3

A Review of Materials and Chemistry for Secondary Batteries

R. Vasant Kumar and Thapanee Sarakonsri

The previous chapter was concerned with the evolution of electrochemical cells, whose material chemistries allow for only one discharge to take place. As described, however in Chapter 1, the most advanced battery chemistries are those which allow for recharge to occur and, hence, the cell can be used multiple times. These reusable cells are therefore termed rechargeable or secondary batteries. To the nonexpert this term is a misnomer in the sense that secondary batteries are not inferior to primary; the coinage of terms relates to history, as it can be seen in the timeline (Table 1.4) presented in Chapter 1.

A secondary battery can effectively be reused many times after it is discharged by applying electrical power to the cell to bring about the reverse reaction. In an electrochemical cell the forward reaction is spontaneous while the reverse reaction that takes place during charging is thermodynamically unfavorable (ΔG = positive). In order to reverse the reaction, the minimum applied potential required can be calculated from $\Delta G = -nFE$. For the Zn/Cu Daniel cell (also known as the gravity or crowfoot cell), described in Chapter 2, the voltage given off upon cell operation is $E^{\circ}_{\text{cell}} = 1.10$ V. Thus, if an electrical potential more than 1.10 V is applied to the cell, the following charging reaction could take place:

$$Zn^{2+}(aq) + Cu(s) \rightarrow Zn(s) + Cu^{2+}(aq)$$

However, as explained previously, the recharging of primary cells is not practiced. After a secondary cell is fully charged, the electrodes attain their original state, allowing the discharge reactions to occur spontaneously and, thus, the cell can be used again to provide energy to an electric appliance/device. This charge–discharge process is referred to as an electrochemical "cycle" and due to irreversibilities in the cell system, it cannot be repeated indefinitely and eventually the cell must be discarded. Reasons leading to the irreversibility of battery reactions are: (i) the formation of an electrochemically inaccessible permanent phase during and/or at the end of discharge, (ii) the loss of mechanical integrity and fracture that result from the buildup stresses that occur during cycling. Due to these irreversible transformations, a capacity loss in occurs during cycling and eventually the capacity tends to zero.

High Energy Density Lithium Batteries. Edited by Katerina E. Aifantis, Stephen A. Hackney, and R. Vasant Kumar
© 2010 WILEY-VCH Verlag GmbH & Co. KGaA, Weinheim
ISBN: 978-3-527-32407-1

Table 3.1 Some typical properties of secondary batteries [1].

System	Nominal cell voltage (V)	Specific energy (Wh/kg)	Cycle life (up to 80% initial capacity)	Charge time (h)	Self-discharge per month (%)
Lead–acid	2	30–50	200–350	8–16	5
Ni–Cd	1.25	45–80	1500	1	20
Ni–MH	1.25	60–120	300–500	2–4	30
Li ion	3.6	110–180	500–1000	2–4	10
Li polymer	3.6	100–130	300–500	2–4	10

Table 3.2 Global battery market forecast for 2010.

Type of battery	Global demand in US$ in billion
Primary	Primary total: 42
Carbon–zinc	8
Alkaline	22
Others	12
Secondary	Secondary total: 54
Lead acid	28
Ni–Cd/Ni–MH and others	12
Li	14

The key properties of the main commercial secondary batteries are shown in Table 3.1, while the commercial demand of these chemistries is shown in Table 3.2.

The secondary battery market is predicted to grow by 20–25% by the end of this decade, driven by the growth in the automotive and e-bikes sectors, the emerging hybrid electrical vehicles (HEVs) as well as the green energy storage markets and the reinforcement of emergency stand-by power in electrical grids and local/regional micro-grids distributing power to customers. An overview of the development of secondary batteries will be given in the present chapter and reference to respective applications will also be made.

3.1
The Lead-Acid Battery

The first secondary (rechargeable) battery system was invented by Raymond Gaston Planté, French physicist, in 1859, for powering the lights in train carriages. Both electrodes consisted of lead, while the electrolyte was acidic, and therefore this prototype rechargeable battery is known as the lead-acid battery. It is also called the SLI battery since it controls the starting, lighting, and ignition of automobiles. The initial versions of these cells were composed of a single unit that

PbO$_2$ electrode   Pb electrode

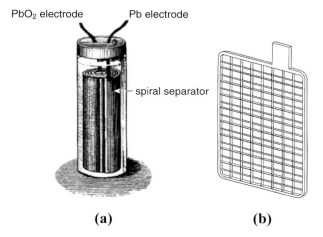

spiral separator

(a)   (b)

Figure 3.1 (a) An illustration of Planté's original lead-acid cell [2]. (b) The pasted plate grid in the later design [3].

contained two lead electrode sheets separated by a spiral rolled coarse cloth spacer, which was later replaced by more stable rubber tapes and strips; the cell was then filled with a sulfuric acid electrolyte solution. With continuous electrochemical cycling the electrolyte corroded the lead producing lead dioxide at one electrode and spongy lead at the other, increasing, therefore, the electrode surface area that could participate in the reaction. As a result the capacity of the battery increased after the first cycle. At that time (in 1859) this cell was also trialed for load-leveling.[1] The original Planté cell is shown in Figure 3.1a.

After continuous research an improved lead-acid cell with increased capacity was developed by Planté's pupil, Camille Alphonse Faure in 1880–82 using a lead foil as the anode and a plate of lead oxide–sulfuric acid (PbO$_2$–H$_2$SO$_4$) pastes pressed on a lead grid lattice as the cathode. This original idea of paste formed cathodes has been more or less retained to the present day. A pasted plate grid is illustrated in Figure 3.1b. Following the Faure design, several improved lead-acid battery configurations were developed, such as the lead–antimony alloy grid that was introduced by Sellon in 1881 which increased the mechanical strength of the electrodes considerably. The lead–calcium alloy grid was designed by H. E. Harring and U. B. Thomas in 1935, resulting in greater safety by reducing corrosion and gassing. The basic architecture of modern lead-acid automotive batteries resembles the Faure design.

Despite the arrival of several other secondary batteries in the last century, the lead-acid battery retains the top spot in the secondary battery industry with a market value approaching \$28 billion per year (Table 3.2), as there are no economic alternatives to automotive starting, lighting, and ignition (SLI) applications.

1) Load-leveling is the method by which electricity is stored during low demand periods (charged at night) for use in high demand periods (discharged during the day), in order to reduce large electrical demand fluctuations.

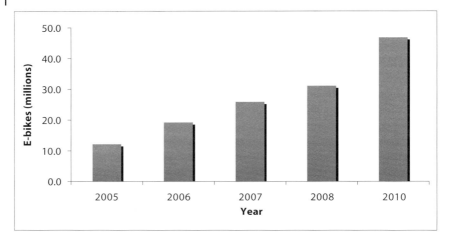

Figure 3.2 Growth of e-bikes in China (note that the number of e-bikes produced were less than 0.1 million in 2000) [4].

Lead-acid batteries are also the power source in over several hundred millions of e-bikes (Figure 3.2) and have the potential to make a major impact in hybrid and electrical vehicles in the near future.

The European Union emission targets (<130 kg CO_2 per km) for vehicles are already driving the increased use of the so-called absorptive glass mat/value regulated lead acid (AGM-vRLA) batteries for micro-hybrid electrical vehicles (micro-HEVs). The small electric motor of the vehicles is powered onboard by a 12–14 V AGM-VRLA battery for immediate restart after stopping in the idle, stop, and go sequence. The engine restart that occurs after each automatic stop results in a large number of high-rate load phases during the battery cycle life. The start-stop cycle in the micro-HEV is expected to decrease battery cycle life marginally; hence the replacement market will rapidly increase in subsequent years [5]. If the battery life is shorter, more batteries will be required for the same number of vehicles. A number of major car manufacturers are planning to convert up to 80% of their new hybrid vehicle fleets into micro-HEVs by 2012, while by 2015 it is predicted that all HEVs will be micro-HEVs. In addition to the motive market, lead-acid cells have an established presence in stand-by emergency power for medical, telecommunication, and memory back-up applications.

3.1.1
Electrochemical Reactions

It has already been noted that the positive electrode mass consists of porous PbO_2, while the negative electrode is made of porous Pb. For both electrodes the aforementioned active materials are pasted on lead-alloy grids and the following reactions take place during discharge

The anode:

$$Pb(s) + SO_4^{2-}(aq) \rightarrow PbSO_4(s) + 2e^- \quad E = 0.35 \text{ V}$$

The cathode:

$$PbO_2(s) + SO_4^{2-}(aq) + 4H_3O^+(aq) + 2e^- \rightarrow PbSO_4(s) + 6H_2O(l) \quad E = 1.70 \text{ V}$$

Overall:

$$Pb(s) + PbO_2(s) + 2SO_4^{2-}(aq) + 4H_3O^+(aq) \rightarrow 2PbSO_4(s) + 6H_2O \quad E = 2.05 \text{ V}$$

The above reactions are simplified versions of the reality, since sulfuric acid is dissociated to bisulfate ion rather than sulfate ions in the acid solution:

$$H_2SO_4 \Leftrightarrow H^+ + HSO_4^-$$

Thus, the anodic, cathodic, and overall reactions involve the bisulfate ions rather than sulfate ions, giving rise to the following overall reaction:

$$Pb(s) + PbO_2(s) + 2H^+ + 2HSO_4^- \Leftrightarrow 2PbSO_4 + 2H_2O$$

The above equation is known as the "theory of double sulfation" and was laid down in 1882 by Gladstone and Tribe. Thermodynamically, Pb should corrode in the acid solution to form lead sulfate while liberating hydrogen gas. Thankfully, this reaction is limited due to the high overpotentials for hydrogen and oxygen evolution on both the lead and lead sulfate. Although this reaction is limited, it is not negligible, however, and can be enhanced by impurities in the spongy active lead, as well as by the presence of alloying elements (such as Sb) in the lead grid. By similar thermodynamic considerations, lead dioxide can also self-discharge by decomposing water to oxygen while converting itself to lead sulfate. The overall water decomposition reaction to gaseous hydrogen and oxygen in the presence of lead and lead dioxide in sulfuric acid is, thus, limited, but not negligible over a period of time. Battery designs must take this into account to minimize water loss and to avoid the explosive hydrogen–oxygen recombination reaction.

3.1.2
Components

The reason that both the anode and cathode active materials are made porous is to increase the high specific active surface over which the electrode reactions can occur, allowing, thus the electrode reactions to take place more efficiently. In particular, the volumetric porosity of the active masses is over 50% at both electrodes. During the reactions sulfuric acid (H_2SO_4) is consumed and lead sulfate ($PbSO_4$) is produced at both electrodes. The discharge capacity is directly increased as a result of increased availability of active components and a decrease in the current density. Initially the electrodes comprise of PbO, which is converted into porous Pb (anode) and porous PbO_2 (cathode) on initial charging. The lead grids provide mechanical support for the electrode pastes and also act as electronic conductors between the electrodes and the load. The combination of a grid and

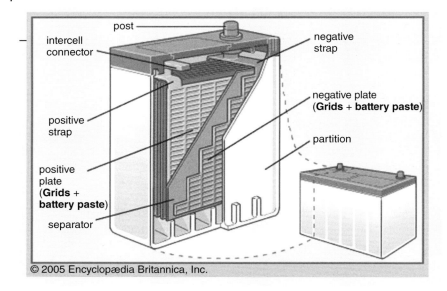

Figure 3.3 A cut-away version of a lead-acid battery [4]; over 80% lead is consumed in making batteries.

the paste on it is referred to as the plate. A separator, usually made of micro-porous polyethylene, is used between the positive and negative plate in order to prevent electrical shorting while the container and the lid are made of polypropylene, which is light, strong, does not embrittle at freezing temperatures, and most importantly it is resistant to attack by the sulfuric acid. A schematic cut-away diagram is shown in Figure 3.3. The specific electrical conductivity of the H_2SO_4 aqueous solution electrolyte in a lead-acid battery is of the order of 0.6 to $0.8\,\Omega^{-1}\,cm^{-1}$, while that of Pb is $0.5 \times 10^5\,\Omega^{-1}\,cm^{-1}$, and that of PbO_2 is $0.7 \times 10^3\,\Omega^{-1}\,cm^{-1}$. Due to sulfation, however, that takes place during discharge, the resistance of the active masses at both electrodes increases. In normal discharge, for the negative active mass, the decrease in conductivity is about 30%, while for the positive active mass, it is around 90%; Despite this large decrease in conductivity, the resistance of the electrolyte is still dominant.

Until two decades ago, in order to achieve rigidity in the grid, it comprised of Pb–10 wt.%Sb. But Sb is prone to dissolution in the acid, resulting in loss of water through dissociation. Thus, in order to maintain the battery, water addition was frequently required. The acid was, hence, contaminated and hydrogen was also generated requiring controlled venting to avoid explosions. As a result the battery life was relatively short at only 2 years.

Modern batteries are made to be maintenance free and need no water addition throughout their life, which is extended to 4–5 years. The amount of Sb is decreased to 1.5 wt%, while hardening of the Pb is achieved using 0.1 to 0.5 wt.% Ca in the Pb alloy. Traditional SLI batteries are also termed flooded batteries as the cell must be filled with water in order to compensate for the loss of water, to hydrogen and oxygen, during electrolysis. The previously mentioned valve regulated lead-acid (VRLA) cells, which dominate the micro-HVE industry, minimize the need to refill with water by suitably changing the composition of the alloy grid. The polyethylene separator is replaced by an absorptive glass mat (AGM) separator, which absorbs the electrolyte, avoiding flooding and slippage, while the hydrogen and oxygen that evolved during electrolysis are allowed to recombine back into H_2O on catalyst plugs, avoiding thus, water loss. At the same time a sealed valve vents gas out only if the internal pressure exceeds a certain critical value. The porous separator can also allow for oxygen to diffuse from the positive PbO_2 electrode to the negative Pb electrode, where it can react with the Pb to form PbO; this prevents the anode from reaching a potential where hydrogen is easily produced. During charging PbO is reduced back to Pb.

However, there are unavoidable limitations of lead-acid batteries, as service life is shortened by (i) corrosion on the positive plate caused by operating the battery at high temperatures above 21 °C or by leaving the battery uncharged. A schematic figure of a corroded grid is shown in Figure 3.4. (ii) Irreversible sulfation on the negative plate occurring when the battery is left unused for a period of time. The product of the discharge reaction which is lead sulfate will irreversibly recrystallize to larger particles. Low surface area and nonconducting lead sulfate particles consequently form during discharge and subsequently block the conductive path needed for recharging; their formation is, therefore, associated with a capacity loss.

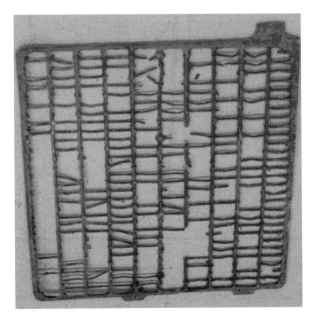

Figure 3.4 A corroded conventional lead grid on the positive plate; taken from [6].

Moreover, lead sulfate has a 37% higher volume than lead or lead alloys which will result in the expansion and deformation of the grid. Appearance of short circuits at the edges of the plates and through the separators also causes sulfation of the plates.

SLI batteries are not designed for deep discharge and they should be kept in an open circuit in order to minimize corrosion, but they should be charged regularly to prevent sulfation. They are designed to contain a large number of relatively thin plates in order to maximize the surface area and current output. Before the automobile is started, a SLI battery is normally in the fully charged state and any lost charge is replaced by the alternator and the battery is returned to its fully charged state. Discharge below 50% will damage the plates permanently. Once the battery is covered in sulfates, the battery is dead and will not respond to recharging.

The latest commercially available lead battery is the bipolar battery design, which comprises of a lead metal anode in a sponged powder form pasted on lead grids and a cathode in a lead dioxide powder form pasted on lead grids; the cell is filled with a 4.5 M H_2SO_4 electrolyte gel instead of liquid. The battery consists of six cells that are placed in series inside the casing such that the positive and negative active materials are applied to each side of the electrode plates and interspersed with insulating separators. Each single cell provides a voltage of 2 V, which makes it a 12 V battery. The greatest challenge is to replace the lead alloy grids with other electronically light materials. By reducing the weight by 50%, a lead-acid battery is capable of achieving an energy density approaching that of Ni–Cd or Ni–MH batteries. To achieve this weight reduction, the following approaches are being trialed: (i) a porous carbon-foam/carbon felt matrix filled with lead deposits; (ii) carbon particles bonded within a polymer film; (iii) conductive titanium oxide particles forming a composite with an epoxy resin; (iv) Ti–SnO$_2$ composite electrodes.

3.1.3
New Components

Recently there was a report on a new invention of the Microcell™ foam by Firefly energy (research and development laboratories of Caterpillar, Inc). The Microcell™ foam is a corrosion-resistant material with superior physical and chemical properties over conventional lead materials. The Microcell™ foam consists of impregnated lead oxide slurry on grids, which is then further converted to lead and a lead dioxide sponge (shown in Figure 3.5) that contains hundreds of spherical microcells. These microcells result in a diffusion path in the micronscale, which enhances the reaction rate (in conventional batteries millimeter diffusion distances take place). The ISL battery employing the Microcell™ is reported to survive in very low temperatures such as −18 °C. The comparison of the Microcell™ foam electrode with a conventional lead electrode is shown in Figure 3.6.

Advances of this new technology introduced by Firefly are as follows:

The conventional lead electrode encounters a volume change of the active material during cycling that is approximately 37%. This new electrode, however, was

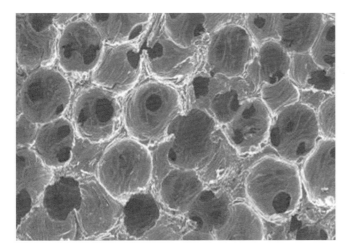

Figure 3.5 SEM image of of Microcell™ foam electrode by Firefly energy; taken from [6].

Figure 3.6 The comparison of a Microcell™ foam electrode (right) and conventional lead electrode (left); taken from [6].

designed to accommodate these volume changes, through its micro-foam structure, as shown in Figure 3.7, indicating a constant volume that does not change during electrochemical cycling. Dispensing with the volume changes inhibits electrode pulverization, and battery failure, due to material failure, can be prevented.

Replacing the conventional negative lead metal-based plate with a Microcell™ composite foam electrode in a 3D architectural technology offers a cycle life that is four times longer than that of conventional lead-acid batteries, due to the fact that sulfation is reversible in 3D cells. The Microcell™ foam electrode material is

Figure 3.7 Microcell™ foams do not experience mechanical damage during electrochemical cycling. (a) An uncycled Microcell™ foam, (b) Microcell™ foam after usage; taken from [6].

Figure 3.8 Oasis battery from Firefly energy; taken from [6].

not reactive in the acid and, hence, does not corrode. Reversible sulfation, un-corroded material, and low self-discharge rates result in longer storage times and longer periods of unused battery to about 3–6 months. The commercial battery of Firefly energy is called the Oasis battery and a schematic diagram is shown in Figure 3.8. The maximum capacity of the lead acid is approximately 170 Wh/kg.

3.2
The Nickel–Cadmium Battery

After the development of the lead-acid battery, it took a whole generation before the second rechargeable battery became commercially available, which is the Ni–Cd battery. Its predecessor was the nickel–iron alkaline battery that was devel-oped by Waldmar Jungner, a Swedish scientist, in 1899. It consisted of a metallic iron negative electrode, a nickel oxide positive electrode, and a potassium hydrox-ide solution with lithium hydroxide acting as the electrolyte. This cell generated a lot of hydrogen gas during the charging process, and therefore required an unsealed battery design. A direct consequence was that leaking could occur easily and therefore a short battery life was encountered. The gravimetric energy of these batteries was 30 Wh/kg, while the cell voltage was 1.4 V and the specific capacity was 21 Ah/kg.

In the same year (1899) Junger invented the nickel–cadmium battery, which was successfully commercialized in 1910 as a pocket-plate battery. Nickel hydroxide mixed with graphite acts as the cathode, while cadmium hydroxide or cadmium oxide mixed with iron or iron compounds acts as the anode; potassium hydroxide solution is used as the electrolyte. On discharge, the $NiO(OH)$ in the positive electrode material changes to $Ni(OH)_2$, while the negative electrode material Cd oxidizes to $Cd(OH)_2$, and water is lost. At the end of every charge some water can be decomposed to hydrogen and oxygen. In some batteries, gas evolution during charge and overcharge is prevented by special design features. For example, the battery is constructed with excess capacity in the Cd electrode; thus, hydrogen is not generated. In order to limit the amount of immobilized electrolyte flat pockets of perforated nickel-plated steel strips were used to hold the active materials; the steel, however, made the battery too heavy for applications in portable devices. The gravimetric energy of these batteries was 27 Wh/kg, while the voltage and the specific capacity were 1.25 V and 20 Ah/kg, respectively. This cell also generated hydrogen gas during charge but in a smaller amount than the nickel–iron battery, and therefore was manufactured in both sealed and open or vented designs. In the sealed design, hydrogen and oxygen gases that were generated at the positive and negative electrodes during charging were recombined back to water. The open design allowed the release of gases and required the periodic addition of water in order to replace the vented gases.

In 1932 the sintered-plate nickel–cadmium battery was invented by Shlecht and Ackermann; two German scientists. The gravimetric energy of these batteries is 30–37 Wh/kg, the voltage is 1.25 V and the specific capacity is 25–31 Ah/kg. The

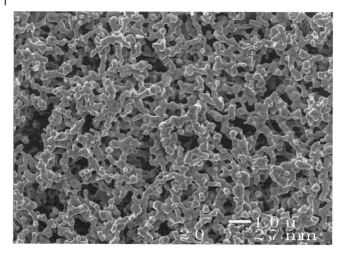

Figure 3.9 Micrograph of a sintered nickel plate; taken from [7].

sintered plate (Figure 3.9) is a high surface area porous substrate impregnated with nickel and cadmium particles, and it is fabricated by a wet-slurry process: a viscous slurry carbonyl nickel is applied on both sides of the nickel-plate grid, which is then sintered at 800–1000 °C. These cells also exist in both vented and sealed designs. The vented design gives a high rate performance which is suitable for engine ignition (aircraft and diesel), but expensive due to the large amount of nickel used. The sealed design is suitable for portable devices such as cameras, phones, computers, and camcorders due to the advantage of being maintenance free (it is not necessary to regularly add water).

The further improved sealed nickel–cadmium battery is a jelly-roll design in which the cathode and anode layers are rolled into a cylindrical shape. This latest commercial design of the nickel–cadmium battery is mostly used in portable devices such as mobile phones, digital cameras, and power tools. The components of this cell are a cadmium metal anode, a nickel oxide hydroxide or nickel (III) oxide (Ni_2O_3) cathode, and a potassium hydroxide (with a small amount of lithium hydroxide) electrolyte solution. The gravimetric energy of these batteries is 40–60 Wh/kg, the voltage is 1.2 V, and the specific capacity is 50 Ah/kg. The modern Ni–Cd configurations are shown in Figure 3.10.

The drawback of all Ni–Cd rechargeable battery systems is that they need to be fully discharged before the charging process should be initiated in order to obtain maximum cell voltage. Otherwise, upon charging the voltage at which discharge ceased will be attained, instead of the maximum voltage possible. This deficiency is referred to as the "memory effect." The memory effect usually is more pronounced as the number of charge–discharge cycles is increased. The effect can be

Figure 3.10 Modern configurations of Ni–Cd batteries [8].

erased in most normal circumstances by a full discharge followed by a full charge. It is reported that the memory effect is connected with the changes in the crystal size involving the reaction between Ni and Cd in the negative electrode material and the large surface area of the sintered nickel support.

3.2.1
Electrochemical Reactions

The reaction at the anode is

$$Cd(s) + 2OH^-(aq) \rightarrow Cd(OH)_2(s) + 2e^- \quad E = 0.82 \text{ V}$$

The reaction at the cathode is

$$2NiO(OH)(s) + 2H_2O(l) + 2e^- \rightarrow 2Ni(OH)_2(s) + 2OH^-(aq) \quad E = 0.48 \text{ V}$$

The overall reaction is

$$Cd(s) + 2NiO(OH)(s) + 2H_2O(l) \rightarrow Cd(OH)_2(s) + 2Ni(OH)_2(s) \quad E = 1.3 \text{ V}$$

The reversible potential for the overall cell reaction for a Ni–Cd battery is 1.3 V and has a higher energy density and a longer life than a Pb–acid battery. For a typical 25% depth of discharge (DOD), the cell can accept recharging for several thousand times. Even at a very high DOD approaching 80%, the cell can be recharged a few hundred times. This is because degradation reactions in the cell are not massive. Over a long period of usage, however, the NiOOH is converted into Ni dioxide hydroxide which is not reversed back into usable active material. A major problem with Ni–Cd batteries has been the use of toxic Cd and using this battery in toys is not appropriate.

3.3
Nickel–Metal Hydride (Ni-MH) Batteries

The Ni–MH batteries began their life as nickel–hydrogen (Ni–H) cells in the 1970s, mainly driven by research of hydrogen storage in metallic alloys. The first Ni–MH batteries were commercially available in 1989 as a variation of the Ni–H cells and their development is associated with the name of the inventor Masahiko Oshitani from the Yuasa Battery Company. He is currently a president and CEO of Blue Energy Co., Ltd. As the name suggests, the cathode in a Ni–MH battery is identical to the one in a Ni–Cd battery, while the anode is a compound of alloys expressed as AB_5 where A is comprisesd of an alloy of rare earths: La, Ce, Nd, and Pr while B is an alloy of Ni, Co, Mn, and Al. Thus, the MH electrode represents the alloy hydride which can reversibly absorb hydrogen during charging and desorb hydrogen into the electrolyte during discharging. The electrochemical reactions during discharge are

at the anode:

$$MH(s) + OH^-(aq) \rightarrow H_2O + e^- + M \quad E = 0.72 \text{ V}$$

at the cathode:

$$NiO(OH)(s) + H_2O(l) + e^- \rightarrow Ni(OH)_2(s) + OH^-(aq) \quad E = 0.48 \text{ V}$$

overall reaction:

$$NiO(OH)(s) + MH \rightarrow Ni(OH)_2(s) + M \quad E_{cell} = 1.2 \text{ V}$$

The electrolyte is an alkaline solution as in a Ni–Cd cell, and most frequently it is KOH (aqueous). The anodic compound is also sometimes made of AB_2, where A is Ti and/or V, and B is Zr (or Ni) with Co, Cr, Fe, and Mn.

A Ni–MH battery has nearly double the energy density of Ni–Cd at 70–80 Wh/kg with a peak power density ranging between 200 and 1000 W/kg. The nominal cell voltage is 1.2 V, giving a specific capacity of 60–70 Ah/kg. The Ni–MH battery is the battery of choice for aerospace applications and is also promising for use in HEV applications. As a result of the higher energy density and the use of nontoxic elements, the Ni–MH battery has replaced the Ni–Cd battery in many applications in low-end electronics, mobile computing, and wireless communications and processing. In general, the Ni–MH battery is less durable than the Ni–Cd battery and can suffer from high self-discharge, poor cycling ability, and lower shelf-life at higher ambient temperatures. The memory effect of Ni–MH is similar to that described for the Ni–Cd battery. It should be emphasized that damage in these cells can occur during charging, and therefore, specially designed chargers or special charging methods need to be used.

In concluding, it is of interest to note that just as foams are the most promising electrodes in the newly designed lead-acid batteries described in the previous section, extensive research is being performed to use open cell Ni foams as cathodes (Figure 3.11) in Ni–MH cells, as they are mechnically more stable. All these studies are under an experimental stage and details can be found in [9].

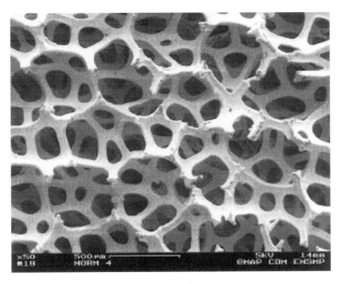

Figure 3.11 Open cell Ni foam, taken from [9].

3.4
Secondary Alkaline Batteries

The competitor of the Ni–Cd battery is the alkaline manganese secondary battery, which was first launched in the market by the Union Carbide Company (USA) in 1946 and is available in D and G sizes (Table 1.1, Chapter 1). This cell consists of a zinc gel (mixed with electrolyte) anode and a cathode comprising manganese dioxide and 10% graphite mixture, which are mixed with aqueous potassium hydroxide (KOH) electrolyte to form a paste; hence the electrolyte and the cathode are one solution. All components are packed in a special hermetically sealed cell.

The Rayovac Company (USA) started producing Rechargeable Alkaline Manganese dioxide-zinc (RAM) batteries cells in 1993 with an energy density of 80 Wh/kg or 220 Wh/dm^3, which is twice that of Ni–Cd batteries. Moreover, self-discharge loss is only 0.01% per year compared to Ni–Cd, which is 1% per year, and most importantly they are environmentally benign and exhibits excellent charge-retention properties over Ni–Cd. RAM cells are currently available in AAA, AA, C, and D sizes, and they are mostly used in toys, games, personal compact disks, cassette players, and tape recorders.

3.4.1
Components

As mentioned, Zn powder mixed with gelled KOH is used as the negative electrode; while MnO_2 mixed with graphite and doped with 10% Bi_2O_3, to allow the reversible process to occur, is used as the positive electrode. $BaSO_4$ and Ag catalyst are also added at the cathode to increase the capacity and the cycle life. The Ag helps to recombine the H_2 and O_2 gases produced by the Zn corrosion. Apart from this, the chemistry and geometry of this battery are similar to that of a primary alkaline battery [5] (see Chapter 2). These cells operate as secondary batteries with a 1.5 V window.

Despite some effort, these cells have not penetrated the market beyond a small segment. In particular, the high internal resistance is restrictive in achieving load currents less than 400 mA, making the cells only adequate for low power demand applications. The cycle life is rather poor at 10 to 50 cycles, depending upon the depth of discharge. Despite the low purchase cost, the cost per cycle is much higher than the well-established secondary cells (Li-ion, Li-polymer, Ni–MH). However, the rechargeable alkaline cell is more economical than the primary alkaline cell. The self-discharge is very low, thus making the alkaline rechargeable battery useful for storing as a stand-by for up to a decade. Currently these cells are produced by the Pierce Energy Battery Company (Canada), and the Alcava range by the Young Poong Corporation (Korea) [6].

3.5
Secondary Lithium Batteries

It did not take long for researchers to realize that lithium-based cells were promising candidates for secondary batteries as well. Due to its low density, lithium has a specific capacity of 3860 Ah/kg in comparison with 260 Ah/kg for Pb. Li is also one of the most electropositive metals, giving rise to very high electrode potentials. Thus, the theoretical energy density achieved by Li is even more favorable in comparison with base metals and transition metals.

The first commercial secondary Li cells were manufactured by Exxon Company (USA) in the 1970s with a $LiTiS_2$ cathode. In 1980, Moli Energy (Canada, now E-One Moly Energy Ltd.) commercialized Li batteries with a $LiMoS_2$ cathode, while a few years later Taridan (Israel) used a $Li_{0.3}MnO_2$ cathode. These three batteries used a Li foil as the anode and various liquid organic electrolytes. The gravimetric energy of the Li–MoS_2 battery is 61 Wh/kg, while that of the Li–MnO_2 battery is 230 Wh/kg. The first two, aforementioned, battery systems are not available presently due to their short life (a few cycles) while the Li–MnO_2 system is not available due to safety issues that resulted from the high reactivity of the metallic lithium anode with the electrolyte; a solution of $LiAsF_6$ in dioxolane solvent had to be used for this cell. Furthermore, use of pure Li as an anode causes safety issues, as Li dendrites are easily formed under a normal rate of charging.

The Nippon Telegraph and Telephone Corporation (NTT), commercialized a AA-size Li battery with a LiV_2O_5 cathode and a Li–Al alloy anode that is used until today. The energy density of these secondary batteries varies from 64 Wh/kg to 135 Wh/kg and their cycle life is between 150 and 400 cycles. The most efficient secondary battery, though, is the lithium-ion battery which was commercially introduced by Sony in 1991 after pioneering research by Goodenough *et al.* [10].

This cell chemistry included a $LiCoO_2$ cathode and a carbon anode, whose potential after being fully charged is 4.2 V. The reversible reaction occurs by the intercalation reaction of lithium ions between the $LiCoO_2$ and carbon frameworks in a nonaqueous liquid organic electrolyte (1 M lithium hexafluorophosphate ($LiPF_6$) in 1:1 weight mixture of EC:DEC (ethylene carbonate:diethyl carbonate)). The face-centered cubic structure of $LiCoO_2$ met the requirement of small volume changes (about 2%) during the Li-insertion and deinsertion that took place upon electrochemical cycling. The delithiated state is believed to maintain its layered structure, which is similar to a hexagonal close-packed structure. Graphite is a suitable anode material since its layered structure with hexagonal vacant sites allows lithium ions to be inserted with a very small volume change taking place (about 11%) and can return back to its initial volume after the delithiation process. The cell potential arising from the voltage difference between delithiated $LiCoO_2$ and lithiated carbon is large, making it a high energy density battery. The gravimetric capacity of this cell is 372 mAh/g or 800 mAh/dm³ and it was shown in Figure 1.12 of Chapter 1 that Li cells have the highest energy density of all secondary cells. However, a safety concern arises from the high reactivity between the lithium and the organic electrolyte in the deep charging state and may result in fire initiation (although this is rare). Use of a gel polymer electrolyte can prevent such risks; This gel electrolyte is used as a very thin film in order to overcome the resistance arising from the low conductivity of the polymer.

Upon full lithium utilization, the cell capacity is 3860 Ah/kg, simply calculated by Faraday's laws. Thus, the actual rated capacity of the cell in Ah is determined by the weight of lithium in the cell. The specific capacity on the other hand takes into account the total mass of the cell and is the ratio of the rated capacity and the mass of the cell. In a general case, the cell mass can be calculated as

$$W_{cell} = w_{Li} f_A + w_{Li} f_C + w_{aux},$$

where w_{Li} is the mass of lithium in the cell, f_A is the multiplier for the anode mass, f_C is the multiplier for the cathode mass, and w_{aux} is the auxiliary mass of the cell, which accounts for approximately 20% of the mass of the active parts of the cell; these parts include the electrolyte, the separator, the connectors, and other remaining components. Thus, as an approximation,

$$W_{cell} \sim 1.2 \, w_{Li} (f_A + f_C)$$

and the specific capacity is given by

$$w_{Li} 3860/1.2 \, w_{Li} (f_A + f_C) = 3860/1.2 \, (f_A + f_C)$$

The specific energy density, which is the energy density divided by the weight of the cell, is the product of the specific capacity and the operating voltage in one full discharge cycle. The values may be quoted for an average voltage or the maximum voltage in the discharge sequence. For a constant current drain, I, the specific energy density, E, can be written as

$$E = \left[I \, w_{Li} \int V \, dt \right] / w_{cell},$$

with the integration being carried out over the full specified discharge time. The specified discharge time is related to a maximum and minimum voltage threshold value to allow for recharging the battery. Outside the threshold values, the battery may attain an irreversible state. Furthermore, the specific energy density is related to the specific capacity, Q, as follows:

$$E = QV_{\text{average}} = w_{\text{Li}}3860/1.2 \ w_{\text{Li}}(f_A + f_C)V_{\text{average}} = 3860/1.2 \ (f_A + f_C)V_{\text{average}}.$$

3.5.1
Lithium-Ion Batteries

As previously mentioned, in a Li-ion cell, both electrodes are based on insertion materials which can insert and deinsert lithium reversibly over many cycles. Graphite and other carbon anode materials can readily achieve this at a potential which is still relatively close to that of pure Li. While the rate of charging, cyclability, and safety are significantly improved with a carbon anode, in comparison with a pure Li, the self-discharge rate is much higher.

A typical Li-ion battery can be schematically represented in Figure 3.12 while the working principle is depicted in Figure 3.13. In addition to the Co-based compounds, the Li cathode can comprise of other chemistries based on Mn, P, Ni, and Fe and various combinations. Table 3.3 provides a comparison between different cathode materials.

In order to maximize the specific energy density, it is desirable to minimize the weight of the cell, while maximizing the ratio of the weight of Li to the weight of the cell. For the Li-ion cell, for example, the theoretical stoichiometric value of the anodic multiplier (f_A) is equal to 10.3 while for the cathode (f_C) it is 25; thus, the maximum theoretical specific energy density for a maximum 4.2 V cell (with dis-

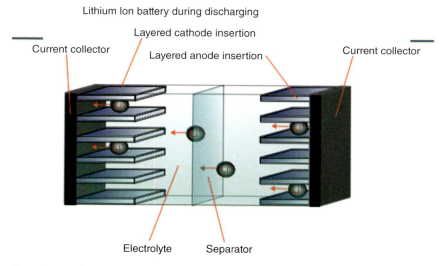

Lithium Ion battery during discharging

Layered cathode insertion

Current collector   Layered anode insertion   Current collector

Electrolyte   Separator

Figure 3.12 Schematic representation of discharge in a Li-ion battery [4].

The working principle for the Li-ion battery

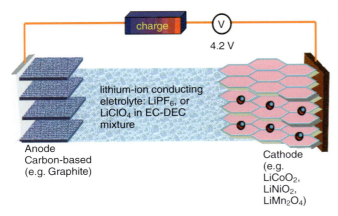

lithium-ion conducting
eletrolyte: LiPF$_6$, or
LiClO$_4$ in EC-DEC
mixture

Anode
Carbon-based
(e.g. Graphite)

Cathode
(e.g.
LiCoO$_2$,
LiNiO$_2$,
LiMn$_2$O$_4$)

Figure 3.13 Working principle of Li movement between the two electrodes [4].

Table 3.3 Comparison of properties of different cathode chemistries [8].

Chemistry	Nominal V (V)	Charge V limit (V)	Charge and discharge C-rates	Energy density (Wh/kg)	Applications	Note
Cobalt	3.60	4.2	1C limit	110–190	Cell phone, cameras, laptops	Since the 1990s it has been most commonly used for portable devices; has a high energy density
Manganese (spinel)	3.7–3.8	4.2	10C cont. 40C pulse	110–120	Power tools, medical equipment	Low internal resistance; offers a high current rate and fast charging but a lower energy density
NCM (nickel–cobalt manganese)	3.7	4.1[a]	~5C cont. 30C pulse	95–130	Power tools, medical equipment	provides a compromise between high current and high capacity.
Phosphate (A 123 system)	3.2–3.3	3.6[a]	35C cont.	95–140	Power tools, medical equipment	New, high current rate, long cycle life, high charge V, increased capacity but shorter cycle life

a) Higher voltage provides more capacity but reduce cycle life.

charge the voltage decreases) is calculated to be between 380 and 460 Wh/kg depending upon whether the weight of the auxiliary components is taken into account or not. In practice, Li availability in both the anode and the cathode is more than halved; thus, the multipliers are typically $f_A > 21$ and $f_C > 50$; the excess value arising from the weight of the binder and other additives.

The operating voltage during discharge is decreased from a maximum value of 4.2 V to a cutoff value of 2.8 V, giving an average value of 3.35 V over the discharge cycle. The practical specific energy density is therefore in the region of 160 Wh/kg for a Li-ion cell. The only practical method for marginally increasing the specific energy density of a Li-ion cell is to decrease the weight of the auxiliary components of the cell further down; if, however, the Li-availability can be increased to the maximum stoichiometric region for both the electrodes, then the specific energy density could be doubled.

The actual specific energy density achieved by a Li-ion battery is in the region of 160 Wh/kg. It is widely believed that with a considerable amount of research and development the maximum specific energy density that can be achieved for a Li-ion cell within the next 5 years could be 250 Wh/kg per cell.

Due to their superior specific energy and energy density, long-lasting small size secondary lithium-ion batteries are suitable for most portable devices such as notebook computers, mobile phones, digital cameras, and most importantly for the future zero emission (or no exhaust) vehicles such as the Tesla Roadster electric car, from Tesla Motors, Inc. (USA), which was released as the first prototype in July 2006. The Tesla Roadster is a two-seat sport car, with an open-top, rear-drive roadster, and 6831 lithium-ion batteries (53 kWh) are used to power it, requiring 3.5 h of charge time. The maximum range for driving is 245 miles. The Tesla cars are reported to currently be in production [11].

To understand the superiority of Li-ion battery powered electric vehicles, a lead-acid battery powered electric car, the Reva car, is described. The Reva car is manufactured by the Reva Electric Car Company, Bangalore, India, since 2001. It is a small three-door hatchback car, which is powered by eight lead-acid batteries (each giving 6 V) that are connected in series to obtain 48 V (9.6 kWh) with 6 h charge time. The maximum range for driving is 50 miles. It is therefore understood why Li-ion batteries are the most promising and sought-after power sources for electric vehicles.

Considerable Research & Development programs have greatly helped the incremental development in the performance of Li-ion batteries (Figure 3.14) accompanied by a massive increase in volume and reduction in cost (Figure 3.15).

It should be noted that Li batteries did not replace in the market, the existing lead-acid or Ni–Cd batteries which maintain their existing market positions. Instead these new rechargeable batteries created the digital and portable device revolution by accelerating the development of portable computers, cellular telephones, and cordless hand-held tools to a degree that was impossible to imagine in the so-called mature market of rechargeable batteries before the 1990s. Both the Ni–MH and the Li-ion cells are produced in the region of 1 billion cells per year each with a market value of approximately $4 billion and $10 billion per year,

Progress in Li-ion Battery Energy Density

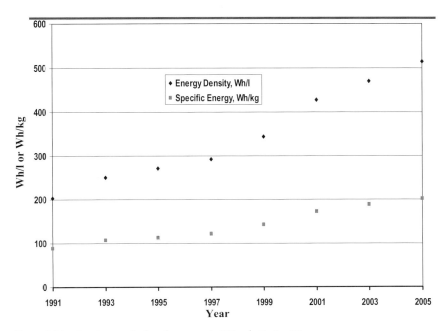

Figure 3.14 Development of performance in Li-ion batteries [4].

respectively. This is a staggering growth considering these cells were not available at all 18 years ago. Ni–MH and Li-ion batteries are not only lighter due to their high specific energy density (Wh/kg) than the Ni–Cd and the Pb–acid rechargeable batteries but are also smaller due to their high volumetric energy density. It is the high values of Wh/kg and Wh/liter that have been the key factors in their rapid growth. It should also be noted that in the early 1990s, the Ni–MH system made good use of the Ni–Cd type of facilities and cutting down on manufacturing lead time.

3.5.2
Li-Polymer Batteries

In 1999 secondary Li batteries were developed that had a similar electrode chemistry as Li ion, but with the liquid/gel Li-ion electrolyte being replaced with a polymer. These batteries were therefore termed Li polymer. Although the same electrodes as those used in Li-ion cells can be employed in Li-polymer cells, other materials may be more efficient.

As the name suggests, a polymer electrolyte can be used to replace a separator soaked in a liquid electrolyte. Use of a dry polymer electrolyte can lead to a higher safety and reduced flammability and offers simplicity with respect to fabrication

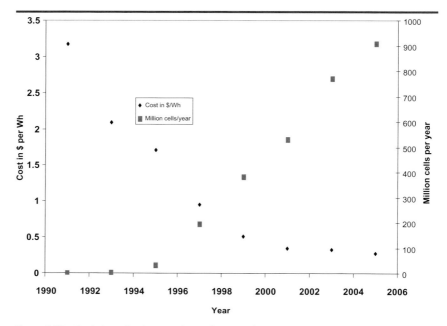

Figure 3.15 Evolution of volume and cost for Li-ion batteries [4].

and geometry. Very thin film type batteries are made possible with a polymer electrolyte. A true polymer electrolyte has low conductivity, thus, offering a higher resistance to charge movement that inhibits the rate of discharge and charging. It should be noted that as a way of compromise, the Li-polymer batteries do not as yet commercially use true polymers but instead use soft polymer solids of the gel-type electrolytes.

3.5.3
Evaluation of Li Battery Materials and Chemistry

As it already shown, Li batteries are the most advanced secondary electrochemical systems and, therefore, the following chapters are dedicated to next generation electrodes and electrolytes for Li batteries. The standard way to test the ability of materials to act as rechargeable Li electrodes is to electrochemically cycle them between the charged and discharged state in order to determine the battery power capability and "lifetime." The lifetime of a rechargeable battery depends on the stability of the voltage:current:capacity relationship as a function of the number of cycles. Such testing can be performed in several ways such as by using flooded and button cells. Button cell testing is more standard and used in most laboratories. Since the majority of the results presented in the remaining chapters were obtained through electrochemical cycling, electrochemical testing will be briefly described here. First of all, the active electrode material needs to be mixed with a binder (e.g., 10% PVDF) and a small amount (~10%) of carbon to make a laminate. Then the laminate is coated onto a Cu foil by a doctor blade; the Cu acts as the current collector. After drying a solvent out in the oven, it is punched out into a circular shape and weighted to obtain the active material used in the cell. The cell assembly is then performed in a dry room using a lithium foil as the counter electrode. An example of a liquid electrolyte chemistry is 1 M LiPF$_6$ in 1:1 EC:DEC. If it is electrolyte candidates that are to be tested, standard electrode materials are used and the electrolyte is substituted. The maximum possible energy in the electrochemical cell is determined by measuring the open circuit voltage (OCV) as a function of the state of discharge. This will act as a benchmark in understanding the amount of internal resistance associated with mass and charge transport. After the OCV measurement, the button cell is put in the cycler (battery test station). The testing procedure starts by discharging (or charging) the cell at a certain current between the maximum and minimum voltages, and once it reaches a minimum voltage, it will automatically charge and so on. The rate of discharge–charge can be varied by varying the current. Then the software gives a plot between voltage and time as shown in Figure 3.16. Multiple plots voltage:time for the charge–discharge cycle can also be obtained. The capacity of the tested cell can then be calculated for each discharge and charge cycle by measuring the area under the curve and dividing by the active material weight to obtain the capacity in mAh/g.

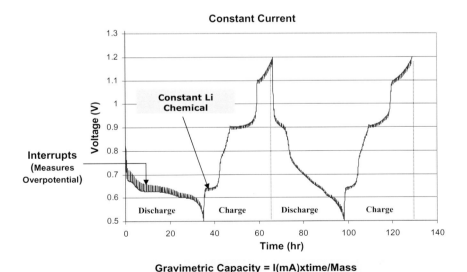

Figure 3.16 Curve obtained from electrochemical cycling.

3.6
Lithium–Sulfur Batteries

The newest type of Li battery invented is the Li–S battery, which has a very high theoretical specific energy density at 2600 Wh/kg (weight of the cell) in comparison with the highly successful and now well-established Li-ion batteries at 460 Wh/kg. This difference arises from the fact that a Li-ion battery uses a carbon anode and a cobalt oxide cathode both of which can at the best support Li in the weight ratio of 1:10 and 1:25, respectively. A Li–S battery is based on a pure Li anode with a 1:1 weight ratio and a sulfur cathode with a theoretical weight ratio of 1:2.3. In practice, both batteries are capable of only partially utilizing the electrodes.

When compared with Li-ion battery development, the Li–S activity has only recently begun and with further research and development, a massive injection of energy density is potentially available. Thus, the starting point for the Li–S battery is the current maximum specific energy density for the Li-ion battery. Currently available Li–S prototypes are already in the 150–220 Wh/kg range and are poised to achieve a target range of 350–500 Wh/kg in the next 5–10 years (see Figure 3.17). It is instructive to compare the rate of progress for the Li-ion battery since its inception in 1991 with that of a speculative scenario for Li–S assuming its practical starting point in the year 2000 (see Figure 3.18).

A Li–S battery can surpass this storage capacity and lead to a new revolution in battery technology. The prize of achieving 500 Wh/kg and 500 Wh/liter is worth the technological, manufacturing, and commercial effort that it is currently undertaken. A Li–S battery can also produce a high power density comparable to that provided by Ni–Cd batteries, which makes it very attractive for high energy-high power applications. But unlike a Ni–Cd battery, a Li–S battery is not known to suffer from memory effect and is very tolerant to overcharging.

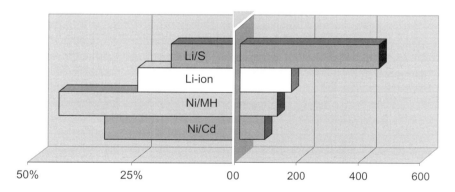

% of Theoretical Energy Density   Practical Energy, Wh/kg

Figure 3.17 Even at a low % of theoretical energy usage, Li/S cells have a high energy density [4].

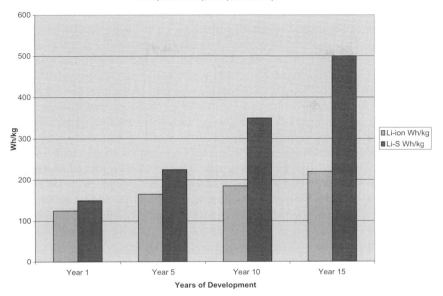

Figure 3.18 Current demonstration of Li–S batteries has achieved over 350 Wh/kg of energy densities; however, the capacity fade with cycling is still being solved [4].

The potential availability of very high energy density puts the Li–S battery in a league of its own, which can become the new standard bearer for another battery revolution in the next 15 years. The following factors provide convincing pointers for a potential exponential growth that the Li–S battery can provide:

- A growing demand for significant weight reduction and/or higher runtime in portable digital, power tool, and medical devices.
- Lower voltage trends of consumer electronic products (2.1 V for a Li–S battery vs. 3.8 V for a Li-ion battery).
- Innovations in wireless, portable devices for digital, hand-held tools and medical applications.
- Growing interest in electrical vehicles (EVs) and HEVs requiring high energy and high power densities.
- Li–S cells can make use of the same manufacturing facilities as Li-ion cells, thus, potentially lowering the development costs.
- The Li–S battery is based upon intrinsically safer materials than Li-ion or Ni–Cd batteries.

Although the Li–S battery has recently become popular, the potential applicability of the Li–S battery as a power source had been mentioned more than two decades ago. Its development though was impeded by the difficulty in the safe use of pure Li as an anode on one hand and even more critically by the electrochemical use of S without encountering charge transfer problems. Furthermore, the formation of Li dendrites on the Li-anode during recharge, and the loss of available S at the cathode, resulted in a rapid fade of the capacity.

Early difficulties did not deter some researchers and work has proceeded in small pockets. The work in the USA (Sion Power), Russia (notably in Ufa Academy of Sciences), UK (Oxis Energy), and Korea (Samsung) is notable. Commercial developments in the Li battery technology, firstly primary cells in the 1980s and later in the rechargeable Li-ion cells in the 1990s, have on one hand led to a decrease in work on the Li–S battery, but on the other have led to new possible development avenues for realizing the potential of a Li–S battery. In particular, considerable research in the arena of Li power sources has (i) directly helped in increasing the knowledge for the safe use of Li, (ii) led to several promising avenues for selecting suitable electrolytes some of which can be modified for the Li–S cells; in fact, the lower the voltage of a Li–S cell the more forgiving it is with respect to electrolyte stability, (iii) resulted in fabrication and processing technologies that can be used to assemble Li–S cells, (iv) opened up electronics/microprocessors for managing battery cycling.

The electrochemical reduction of S in a Li–S cell is a complicated multistep process from $S_8 \rightarrow Li_2S_8 \rightarrow Li_2S_6 \rightarrow Li_2S_4 \rightarrow Li_2S_3 \rightarrow Li_2S_2 \rightarrow Li_2S$ (Figure 3.19). The most reduced state of sulfur Li_2S cannot be accessed in the electrochemical cycle due to insolubility in the liquid electrolytes. Li_2S_2 is also insoluble, but it is chemically leached into the electrolyte solution by reacting with higher polysulfides, which in turn are all soluble. In fact, the choice of the electrolyte is normally made to ensure that it can serve as a solvent for most of the products. This, however,

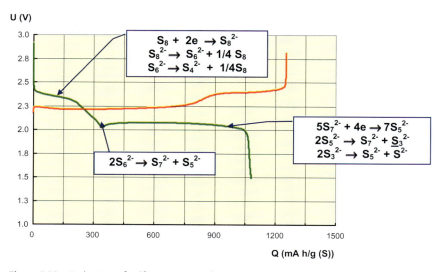

Figure 3.19 Reduction of sulfur in a Li–S cell [4].

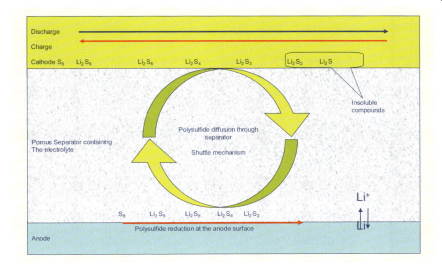

Figure 3.20 Shuttle mechanism [4].

leads to other serious problems. The plateau of 2.1 V is due to two-phase equilibria between a solid lithium sulfide phase and the dissolved lithium sulfide or possibly between two solid phases. A solid Li sulfide phase can passivate the C particles (added for conductivity) in the electrode, thus preventing charge transfer during charging and subsequent discharge.

A decrease in sulfur capacity during cycling also arises due to the irreversible dissolution of polysulfides in the electrolyte which is shuttled away toward the anode and parasitic reactions take place directly with Li to form the more oxidized form of the polysulfide, and also due to the formation of electrochemically inactive Li_2S. Parasitic reactions of soluble polysulfides among themselves also lead to significant self-discharge. The shuttle mechanism (Figure 3.20), however, is beneficial for offering overcharge protection during charging. There has been considerable effort in recent years to develop a suitable Li–S energy storage system capable of achieving effective cycling of a sulfur electrode. The challenge is to achieve the required 1000-cycle life with a self-discharge that is less than 15% per month.

In concluding this section, it should be noted that many companies and researchers rejected the idea of cobalt oxide as an electrode in the late 1980s before Asahi Chemicals and Sony, both in Japan, pioneered the Li-ion revolution. It is, therefore, possible that pioneering groups in Li-battery research may devote their research toward the Li–S cell, allowing hence, its commercialization. Given the price advantage for the Li–S battery over the Li-ion battery (a potential factor of 0.5) it is also conceivable that a Li–S battery can replace Li-ion batteries in some of their current applications.

3.7
Conclusions

The purpose of this chapter was to provide an overview of the evolution of secondary batteries. As was seen, the most promising electrochemical cells that are currently used commercially are Li ion and Li polymer. Therefore, the subsequent chapters of this book will describe the material candidates to be used as cathodes, anodes, and electrolytes, in next generation Li batteries. The Li–S cells are not included in the remainder of the book as they are still quite away from commercialization.

References

1 Gaston, P. (1859) The Storage of Electrical Energy, contributed by Osmania University, Paul Bedford, India. http://www.archive.org/details/storageofelectri031249mbp (accessed 6 November 2007).

2 DOE (1995) Primer on lead-acid storage battery, DOE handbook, DOE-HDBK-1084-95. http://www.hss.energy.gov/nuclearsafety/ns/techstds/standard/hdbk1084/hdbk1084.pdf (accessed 8 February 2010).

3 http://www.engineersedge.com/battery/negative_positive_plate_construction.htm, Copyright 2000–2010

4 University of Cambridge (2005) DoITPoMS Teaching and Learning Packages. http://www.doitpoms.ac.uk/tlplib/batteries/index.php (accessed 15 November 2007).

5 Schaeck, S., Stoermer, A.O., and Hockgeiger, E. (2009) *J. Power Sources*, **190**, 173.

6 Firefly Technology, Technology of our Innovation (2008) Firefly's composite foam-base battery technology. www.fireflyenergy.com (accessed 2 December 2008).

7 Linden, D. and Reddy, T.B. (2002) *Handbook of Batteries*, 3rd edn, McGraw-Hill, New York.

8 Buchmann, I. (2003) http://www.batteryuniversity.com/index.htm/ (accessed 15 October 2005).

9 Badiche, X., Forest, S., Guibert, T., Bienvenu, Y., Bartout, J.-D., Ienny, P., Croset, M., and Bernet, H. (2000) *Mat. Sci. Eng. A*, **289**, 276.

10 Mizushima, K., Jones, P.C., Wiseman, P.J., and Goodenough, J.B. (1980) *Mater. Res. Bull.*, **15**, 783.

11 http://www.teslamotors.com/buy/buyshowroom.php

12 Pistoia, G. (2005) *Batteries for Portable Devices*, Elsevier Science B.V., Amsterdam.

13 Besenhard, J.O. (1999) *Handbook of Battery Materials*, Wiley-VCH Verlag GmbH, Weinheim.

14 van Schalkwijk, W.A., and Scrosati, B. (2002) *Advances in Lithium-Ion Batteries*, Kluwer Academic Publishers, Massachusetts.

15 Crompton, T.R. (2000) *Battery Reference Book*, 3rd edn, Newnes, Oxford.

16 Huang, C.-F. (2005) On lithium cobalt oxide lithium-ion battery cathode materials surface modification and battery properties, Master's Thesis, URN93324032. National Central University, Taiwan.

4
Current and Potential Applications of Secondary Li Batteries

Katerina E. Aifantis and Stephen A. Hackney

As described in the previous chapter, secondary Li batteries are the most rapidly growing high energy storage devices in industry. The advantages of Li cells over other rechargeable electrochemical cells are: they do not exhibit a memory effect, they provide a higher energy density per unit mass and their self-discharge rate is less than half compared to the next best available solutions such as NiCd and NiMH cells. Furthermore, the lifetime of current commercial Li batteries is over 1000 cycles, while their shelf-life is more than 10 years [1]. On the negative side, the lithium-ion battery materials are more fragile and require a protection circuit to ensure safe operation; however, the recently developed Li-polymer cells are easier to handle with safety [2]. The purpose of the present section is to give an overview of the main, current and future, applications that secondary Li batteries have in our society, so as to motivate the reader to focus on the chapters to follow about improving Li battery technology. After the more general application information is provided, detailed information will be given concerning how secondary lithium batteries power electric vehicles.

4.1
Portable Electronic Devices

Presently, the main application of rechargeable Li batteries is in portable electronic devices, such as cellular phones, digital cameras, global positioning system (GPS) devices, and laptop computers. As the key feature of these devices is portability, in addition to energy efficiency requirements, the size and weight of the Li cells are a main concern, as in some cases the Li batteries consumed most of the weight of the device.

Cell phones have evolved from ordinary phones, to digital cameras, MP3 players, and wireless internet access devices. In order to accommodate all these features, more advanced Li-ion cells than those used in conventional cell phones had to be developed, as the energy demand is much higher in these new multifunction mobile phones. Continuous research for improving Li batteries is therefore mandatory and as essential as microelectronic advancement. To illustrate the

High Energy Density Lithium Batteries. Edited by Katerina E. Aifantis, Stephen A. Hackney, and R. Vasant Kumar
© 2010 WILEY-VCH Verlag GmbH & Co. KGaA, Weinheim
ISBN: 978-3-527-32407-1

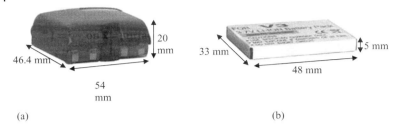

Figure 4.1 (a) 680 mAh (13.51 Ah/l) Li-ion battery for the Motorola StarTac commercially available since 2000. (b) 900 mAh (1136.36 Ah/l) Li-ion battery used in the Motorola Razor V3 model; taken from [3].

significance of battery capacity in device performance the following example is given: the 2000 model of the Motorola Startac used a 680 mAh Li-ion cell which allowed for a talk time of approximately 4 h and a wait time of about 75 h. A more recent Motorola model, the Razor V3 employs a higher capacity Li-ion cell of 900 mAh, which allows for about 7 h of talk time and a standby time of 290 h [3]. The 900 mAh Li battery does not only allow for longer usage times of the phone before recharging is necessary, but it is smaller in size and therefore allows for a lighter and smaller cell phone. This significant size reduction of the aforementioned Li batteries is shown in Figure 4.1.

The other most popular application of Li batteries is in laptop computers. Although Li-ion cells are the most widely used, Li-polymer cells are starting to be preferred for some laptops and other electronic applications. Although they are referred to as Li polymer, they are in fact Li-ion polymer as the electrolyte does not comprise of a pure solid polymer but Li salts and/or gels are added to the polymer to increase conductivity. Pure polymer electrolytes offer a very low conductivity and cannot be used in any application in their present form; current research in polymer electrolytes will be described in detail in Chapter 7. The drawback of commercially available Li-polymer cells is that they are more expensive and require more time to manufacture. The advantages though are that they can be shaped with greater ease so as to occupy more efficiently the available space in the device, they are resistant to damage and there is less danger for flammability [2]. These Li-polymer cells have found their initial applications in the computer industry. Apple was able to develop a very thin laptop, the Mac Book Air (1.94 cm thick), by using an integrated 37 Wh Li polymer battery that allows for a maximum battery performance setting of approximately 4 h [4]. However, the space saved in thickness is gained in length and width as this laptop is 32.5 cm long and 22.7 cm wide.

4.2
Hybrid and Electric Vehicles

Although commercial secondary Li batteries cover the needs of the portable electronic industry satisfactorily, the future of electric vehicles depends on the further development of Li battery technology. In Chapter 3, examples of hybrid and electric vehicles were given and the superiority of Li-powered vehicles was shown. Here, more details will be given concerning this industrial endeavor.

Hybrid electric vehicles (HEVs) are defined as vehicles that use two distinct power sources to operate the vehicle; namely a combustion engine system together with an electric propulsion system. Depending on the primary power source during operation, HEVs can be categorized as parallel HEVs, electric-assisted or mild HEVs and plug-in-HEVs (PEVs).

Parallel HEVs use the electric motor during low demand traveling (low speed, low acceleration) and the internal combustion engine is activated when there is a higher energy demand to move the car.

Electric-assisted or mild HEVs use the electric motor only during heavy operation demands, for example when a high acceleration is needed.

When referring to HEVs the main category that comes to mind is the PEV, due to the fact that they can be recharged from a regular electric plug or charged on the go from the engine. These vehicles depend mainly on the electric motor that is powered from the Li battery energy storage and use the gasoline engine in case of emergency to maintain operation during discharge of the lithium battery.

HEVs, therefore, provide the same performance as conventional vehicles, but through a significantly reduced dependence on fuel. Their environmental friendliness is not only attributed to the consumption of less fuel, but also to the lower carbon emissions that are achieved through the need for smaller internal combustion engines and the use of energy saving technologies such as regenerative braking, by which energy is captured during the braking of a HEV in order to recharge the battery. Moreover, there is a significant noise reduction in HEVs in comparison to the internal combustion vehicles.

Electric vehicles are the more evolved version of HEVs as they operate solely on an electric propulsion system and do not require any form of fuel. Although commercially available Li-ion cells are not the most appropriate for EVs, a commercial EV, the Tesla Roadster, was manufactured by enabling elaborate power-control and monitoring electronics for the battery cells. The battery is a custom microprocessor-controlled lithium-ion battery comprising of 6831 individual cells and weighs 992 lbs. A full charge takes about 3.5 h using the Tesla Motor High Power Connector, which is plugged in a normal household plug [5]. The Tesla Roadster can accelerate from 0 to 60 mph in less than 4 s and has an electronically limited top speed of 125 mph. The range of the vehicle under a full charge is about 220 miles (based on EPA combined city/highway cycle). The battery life is 5 years or 100 000 miles.

In order to develop superior EVs, industry has turned towards using other materials chemistries for the Li battery electrodes. $LiCoO_2$ electrodes used in commercial Li-ion cells wear out after a couple of years, and they become unstable if overheated or overcharged. So, despite these cells being energy-storage efficient, their tendency to exhibit thermal runaway has led the industry to employ alternative cathodes for EV applications. In particular, Co-based cathodes experience an oxidation, upon charging, of Co^{3+} to unstable Co^{4+}. Phosphate-based cathodes, however, undergo oxidation of Fe^{2+} to the stable Fe^{3+} resulting in a safer, fault tolerant chemistry [1]. Therefore, $LiFePO_4$ cathodes are predicted as very promising cathodes to use in EVs, as they not only provide the required capacity and have a lifetime of thousands of cycles, but they also improve the safety during charging. This is important for EVs where safety issues and power capacity are more important than energy capacity. In this connection it should be noted that Li-ion batteries operate best and safest within a charge state window; hence engineers can exploit this fact by charging to a percentage of the maximum capacity. It follows, therefore, that the technology for monitoring the charge status of the Li cells is crucial for the safety and the operation of the EV. However, it is still in the developing stages as it is more difficult than that required to monitor the gas in the tank.

Despite the fact that EVs are zero emission vehicles (the only emission is thermal energy), concerns have risen about the use of electricity for charging the Li batteries that power them. In the USA, 49.7% of the nation's electricity is generated by the burning of coal. According to the Energy Information Administration, for every gallon of gas used by an internal combustion engine, 19.564 lbs of CO_2 are produced. A recent analysis by the Automotive Testing and Development Services found that for every 100 miles of travel, a Tesla Roadster [5], which is an electric high performance vehicle, needs to be recharged with 31 kWh of electricity. Generating a kilowatt hour of electricity produces an average of 1.55 lbs of CO_2. These results will vary according to the use of coal as a main source for electricity per state. In high coal-dependent states the carbon emissions for the Tesla Roadster recharge may be as high as the carbon emissions from an internal combustion engine car. In the technical analysis of the Pacific Northwest National Laboratory [6], it is pointed out that there is plenty of surplus electricity that goes unused at night and this might change with the commercialization of the electric cars without significantly affecting the carbon emissions. Nevertheless, a significant difference in CO_2 emissions between traditional vehicles and EVs exists in states where renewable energy sources are the main electricity source as opposed to coal. In order, therefore, to fully take advantage of the preferred environmental features of EVs, renewable energy source technologies such as solar cells and wind mills must be advanced and reinforced. For example, optional solar cells may be installed on the rooftop of the vehicle to provide additional recharging for the batteries during sunlight.

In concluding, it should be noted that a recent study by the Electric Power Research Institute and the Natural Resources Defense Council suggests that electric vehicles could eliminate billions of tons of greenhouse-gas emissions between 2010 and 2050. A study by General Electric indicates that if half the vehicles on the road in 2030 are electric powered, petroleum consumption in the United States will shrink by six million barrels a day.

Detailed information about engineering secondary Li batteries for EV applications will be given in Section 4.4.

4.3
Medical Applications

4.3.1
Heart Pacemakers

Another area that promises significant applications for next generation recharge-able Li batteries is that of biomedical implantable devices such as pacemakers. Pacemakers comprise of a battery power source which creates electric impulses (periodic electric stimulations) that are carried to appropriate parts of the body through electrodes.

The initial application of pacemakers was to regulate the beating of the heart (treat cardiac arrhythmia disorders) by delivering the electrical impulses to the heart muscles. The first fully implanted pacemaker procedure took place in 1957 at the Karolinska University Hospital in Sweden by Åke Senning and the pace-maker was powered by a secondary Ni–Cd battery, designed by Rune Elmqvist. The Ni–Cd cell, however, was replaced by primary batteries, since the recharging process was not very efficient, as it was performed externally through an appropri-ate vest that provided electromagnetic waves [7]. The primary cells used, thereafter, were toxic, such as the ceramic plutonium oxide battery and the mercury battery.

The real breakthrough in implantable cardiac pacemakers took place in 1972 when the primary Li–I battery was invented by Wilson Greatbatch and his team. This type of battery has a life span of over 10 years and even today is the power source for many cardiac pacemakers. In addition to its long life time the Li–I cell has a very low self-discharge rate, which results in a long shelf-life until the pace-maker is implanted; it also has a stable voltage and it consumes its energy in a gradual and predictable manner. This is important for the follow-ups of patients, where the status of the battery has to be checked and the replacement needs to be programmed. Moreover, the battery produces no toxic waste or gas and is easily encased in a noncorrosive material. An additional significant benefit is that the size of these devices was reduced because of the higher capacity of the lithium batteries used. Lithium carbon monofluoride (CF_x) cells have recently been reported to offer higher energy density and pulses at currents above 20 mA [8], which is slightly better than today's competing batteries. Moreover, the whole battery system is encased with a titanium casing, allowing a 50% reduction in weight over the same size Li–I battery, without any compromise in the safety [8].

Modern heart pacemakers consist of a primary Li battery power source, elec-trodes which are attached on the myocardium of the heart, a sensing amplifier which processes the electrical manifestation of naturally occurring heart beats as sensed by the electrodes, and the computer logic that delivers the pacing impulse to the electrodes. Most commonly, the pacemaker (with the exception of the elec-trodes that are placed on the myocardium) is placed below the subcutaneous fat of the chest wall, above the muscles and bones.

Heart pacemakers have evolved to being able to treat additional heart problems, such as those consisting of multiple electrodes that can stimulate different positions within the heart to improve synchronization of the lower chambers of the heart. Furthermore, modern pacemakers can be programmed externally, once they are implanted, allowing the cardiologist to adjust the pace according to the individual patient needs. Some pacemakers are dual functional as they can also act as a defibrillator.

4.3.2
Neurological Pacemakers

During the past decade pacemakers have found significant applications in neurology as well. Neurological disorders result from chemical imbalances that cause parts of the brain to send out the wrong electric signals. By implanting a pacemaker in the upper chest, electrodes can carry out the proper electric impulses to the dysfunctional parts of the brain and interfere with disordered neural activity, treating therefore various incurable otherwise disorders, such as severe depression, tremor, Parkinson's disease symptoms, and chronic pain.

This treatment is called deep brain stimulation (DBS). The components of DBS are: the implanted pulse generator (IPG), the lead, and the extension. The IPG consists of the primary Li battery, which is encased in titanium, and generates the electric pulses; it is implanted below the clavicle or in some cases at the abdomen [9]. The lead is a coiled wire insulated in polyurethane and has four Pt–Ir electrodes for carrying impulses to the damaged region; they are placed, therefore, directly in the brain by drilling a hole in the skull. The extension is an insulated wire that connects the lead to the IPG; it therefore runs from the head, down the side of the neck, behind the ear to the IPG. The advantages of DBS are that the electric pulses can be calibrated externally depending on the patient's needs and if side-effects appear during operation they completely disappear by turning off the battery. The underlying mechanisms by which DBS functions are not understood; however, it is documented that it can change brain activity in a controlled manner and once the battery is switched off the symptoms resume [10]. Furthermore, there can occur complications such as brain bleeding during the lead implantation, however, this is not common.

The USA Food and Drug Administration (FDA) allowed the use of DBS for the first time in 1997 [11] for the treatment of tremor, and has proven to be very efficient; only 40% of tremor patients can see improvements through conventional medicine. Another area in which DBS has potential is depression. It has been observed that 10% of depression patients see no improvement by using antidepressant medicines, psychotherapy, not even electroshock therapy. An experimental study in 2005 showed that four out of six such patients saw great improvement with DBS [12]. DBS is also used in treating chronic pain, in which cases the pacemaker can be placed on the spine. FDA approval for Parkinson's disease treatment was given in 2002 [11]. Other areas in which it is believed that DBS could help is epilepsy [13] and Tourette syndrome [14].

The most miraculous perhaps application of DBS was reported in 2007 [15]. A 38-year-old male had been in a vegetative state, after his skull was severely damaged during a robbery and he was left for dead. After extensive brain surgery he survived but the doctors gave zero chances of recovery. After 6 years of being in a vegetative state, neurosurgeon Dr. Ali Rezai and his team at the Cleveland Clinic Foundation were able to activate parts of the patient's brain through DBS, by sending appropriate electrical impulses. As a result the patient was able to regain his consciousness, recognize and talk with his family, and perform basic functions by himself such as eating.

Although all medical implantable devices make use of primary Li batteries, instead of secondary Li cells, the aforementioned medical applications were summarized as they illustrate the unique possibilities that Li-powered devices offer in treating incurable, otherwise, diseases. Nonetheless, the current primary Li battery technology is considered to be deficient for DBS applications because of short battery life. This deficiency could be alleviated if the primary battery was replaced with secondary batteries. Unfortunately, current commercial secondary Li batteries cannot be used in medical implantable devices due to safety issues. Therefore, it is of great significance that new material chemistries for rechargeable Li batteries are developed, not only for economic and environmental reasons, but also for improving directly the quality of human life, as it will allow for great improvements in pacemaker applications in medicine and the replacement operations will be diminished.

4.4
Application of Secondary Li Ion Battery Systems in Vehicle Technology

Now that a summary of the various applications in which Li batteries are and can be employed in the future has been given, a more detailed description of how they are used to operate electric vehicles will be given.

The use of a battery in any electrical machine application is limited by the non-ideal behavior of the battery voltage. As reviewed in several chapters here, the open circuit voltage (OCV) is a function of the state of charge with the OCV decreasing as the state of charge decreases. Since the power is the product of current with voltage, a continuous increase in current from the battery would be required as the state of charge decreases. It is, therefore, unfortunate that by increasing the battery current a further decrease in the battery voltage is obtained due to the charge transfer and concentration overpotentials. As a result, the power capability of the battery is degraded to the point where the power requirements of the electrical device may not be met even at a high state of charge. This means that single electrochemical cells must be combined with one another in specific ways to meet application power requirements. As an example, consider that a Li-ion cell operates at a voltage of ~3.5 V. Operation of a small Li-ion cell limited to 1 A of current gives a power output of 3.5 W. The power goals for battery applications in a hybrid vehicle are between 20 000 and 35 000 W [16]. The application of a battery cell to power a hybrid vehicle means that the current and voltage must be scaled up significantly from that found in handheld applications. The power output of an individual electrochemical cell is limited by the diffusive flux of ions through the separator and the electrode material and the thermodynamic driving force associated with this diffusion process. If we look at Eqs. (1.12)–(1.29) it is apparent that the exchange current density and the transport current density are limiting terms for the current. However, as described in Chapter 1 these transport resistances are area-specific terms, which means that simply increasing the electrolyte–electrode interface area will increase the current capability of the cell. It is possible to increase the electrode area by controlling the electrode geometry of the separator–electrode interface, but manufacturing constraints and vehicle geometry constraints may limit the cell separator–electrode interface area. The electrode–electrolyte interface area can also be increased by varying the porosity of the electrodes. Electrode porosity control is achieved through varying the particle size of the active electrode material, the amount of binding compound holding together the active particles and the manufacturing parameters for forming the electrode. Another approach for influencing the internal resistance of an individual cell in order to improve power capability is to increase the effective cross section of the battery by linking individual cells in parallel. This parallel linkage is illustrated in Figure 1.2a for two individual cells. In terms of cell chemistry for the case where the battery properties (state of charge (SOC), OCV and internal resistance) among all the cells are equivalent, the ion transport in each cell occurs independently of the other cells, but the current flow is added together at the junction of the three leads. In electrical terms, this means that the battery terminal voltage is the same as the voltage at each individual cell and the current at the battery terminal is the sum of the individual cells.

It may not, however, be possible to meet the needs of high power applications using only parallel cell linkage. Although the parallel linkage will increase the battery current available to the device with a greater percentage of utilization of the OCV, meeting power demands with very high currents at a relatively low voltage can be problematic. The power loss due to the resistance is proportional to the square of the current. This means that high current solutions will likely result in large inefficiencies due to power dissipation at the electrical machine. This problem is discussed below with specific application to an electric motor. Because of the power dissipation problem associated with a high current, it may be necessary to engineer battery cell clusters, or modules, to deliver a high voltage, as well as, a high current to meet the power requirements of a hybrid vehicle application.

According to Table 4.1, the US Department of Energy (DOE) voltage goals are between ~400 V and 200 V. The effective voltage of the cell chemistry is increased by linking cells in series, with the positive electrode of each cell electrically linked to the negative electrode of the adjacent cell. The voltage of this "string" of cells is measured across the electrodes at the two ends of the string, as shown in Figure 1.2b. In this case, the chemical reaction in each individual cell is linked to the chemical reaction in all the other cells by the electron transfer from a particular negative electrode to the positive electrode in the adjacent cell. It is only the terminal cells at the end of the string that pass an electron to the external device, although the driving force for this transfer is the sum of the free energies for the

Table 4.1 FreedomCAR energy storage system performance goals for power-assist hybrid electric vehicles (November 2002); taken from Ref [16].

Characteristics	Units	Power-assist (minimum)	Power-assist (maximum)
Pulse discharge power (10s)	kW	25	40
Peak regenerative pulse power (10s)	kW	20 (55-Wh pulse)	35 (97-Wh pulse)
Total available energy (over DOD range where power goals are met)	kWh	0.3 (at $C_1/1$ rate)	0.5 (at $C_1/1$ rate)
Minimum round-trip energy efficiency	%	90 (25-Wh cycle)	90 (50-Wh cycle)
Cold cranking power at $-30\,°C$ (three 2-s pulses, 10-s rests between)	kW	5	7
Cycle Life, for specified SOC increments	Cycles	300 000 25-Wh cycles (7.5 MWh)	300 000 50-Wh cycles (15 MWh)
Calendar Life	Years	15	15
Maximum weight	kg	40	60
Maximum volume	1	32	45
Operating voltage limits	Vdc	max $\leq$ 400 min $\geq$ ($0.55 \times V_{max}$)	max $\leq$ 400 min $\geq$ ($0.55 \times V_{max}$)
Maximum allowable self-discharge rate	Wh/ day	50	50

reactions in all of the cells in the "string." This means that the voltages of all the cells in the string add, providing a technique to engineer the voltage of a battery consisting of many individual cells.

The use of individual cells connected in parallel and series to create battery modules allows the tailoring of the cell chemistry for high power applications. The consideration of the performance of these types of modules for a given application can be carried out by utilizing an equivalent circuit approach. Equivalent circuits applied to simulation of battery performance often consist of an ideal (constant) voltage source linked in parallel with electrical components such as resistors and capacitors. The resistors and capacitors linked with the constant voltage source simulate the variation in the battery voltage with the SOC, the current, the time integrated current, and the time derivative of the current. This type of equivalent circuit battery model satisfies all of the standard rules for electrical circuits such as Ohm's law, Kirchoff's law, etc. Because the battery can be represented as a circuit, it is in a convenient form to link with other circuit models used in the design of electric vehicles, such as the electrical motor and electrical control system circuit models. The equivalent circuit models then allow for the introduction of battery performance within the application design process. A simple, but effective

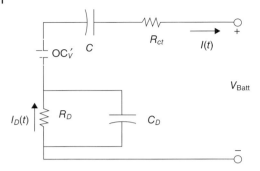

Figure 4.2 FreedomCAR equivalent circuit model for nonideal power source behavior. The electrical components shown simulate the current–voltage relationship in a battery with internal impedance associated with diffusion, charge transfer and OCV decrease associated with SOC. Adapted from [16].

Thevenin equivalent circuit utilized for batteries is presented in the FreedomCAR test manual (DOE) and consists of the circuit elements shown in Figure 4.2. The perfect voltage source having the maximum value (OCV_{max}) for the OCV at a SOC of 1 is in series with an initially uncharged capacitor (of capacitance C), which simulates the fall in the OCV with decreasing SOC. Also, the perfect voltage source is in series with a time independent resistor (R_{ct}) simulating the charge transfer overpotential, and in series with a parallel RC (R_D and C_D) component simulating the concentration overpotential (Warberg impedance). The controlling equations for the battery terminal voltage (V_{Batt}) response to a time-dependent current $I(t)$ prescribed by a function $f(t)$ are then

$$I(t) = f(t) \tag{4.1}$$

with the time-dependent $OCV(t)$ determined by the integrated current as

$$OCV(t) = -\frac{1}{C}\int_0^t I(\tau)d\tau + OCV_{max} \tag{4.2}$$

Unfortunately, the OCV is not the observed battery voltage under practical application. The internal resistances to charge flow within the battery, R_{ct} and R_D discussed above, cause an energy dissipation resulting in a deviation from the OCV. The magnitude of the deviation of the battery terminal voltage from the OCV is obtained by subtracting the time-dependent current through the charge transfer equivalent resistor and the product of the time-dependent diffusion impedance resistor giving

$$V_{Batt}(t) = OCV(t) - [I(t)R_{ct}] - I_D(t)R_D \tag{4.3}$$

The $I(t)$ is already specified by Eq. (4.1), but $I_D(t)$ must be determined from

$$\frac{dI_D}{dt} = \frac{I(t) - I_D}{R_D C_D} \tag{4.4}$$

A solution for $I_D(t)$ from Eq. (4.4) is

$$I_D(t) = e^{\frac{-t}{R_D C_D}} \int_0^t \frac{I(\tau)}{R_D C_D} e^{\frac{\tau}{R_D C_D}} d\tau \tag{4.5}$$

where $I_D(t)$ is the current passing through the resistor in the RC component. Figure 4.3a examines the battery voltage, V_{Batt}, response from Eq. (4.3) to a prescribed sinusoidal, alternating charge–discharge battery current for the case where the integrated current is always small relative to the maximum battery capacity. The expected voltage deviation from the OCV is observed, along with the expected phase difference between the voltage response and the current. The voltage deviation from the OCV is due to the charge transfer and concentration overpotentials, while the phase difference is due to the time dependence in the development of the concentration overpotential; this concentration overpotential is also often referred to as the polarization (therefore $I_D(t)$ is also called the polarization current). This same phase difference between current and voltage can be observed between $I(t)$ and $I_D(t)$ in Figure 4.3b. As the $I_D(t)$ is shunted through the R_D component in the equivalent circuit, it leads to an additional voltage drop from the constant voltage source, which is the concentration overpotential. As a further demonstration of predicted behavior from this type of model, the degradation of power capability due to multiple current pulses is examined in Figure 4.3c. As shown in Figure 4.3c, there is a sudden drop in battery voltage from the OCV at the initiation of the current pulse followed by a slow degradation in battery voltage during the current pulse. The sudden drop in the voltage is associated with the charge transfer overpotential, while the slow degradation in the voltage is due to the time dependence in the development of the concentration overpotential. The multiple pulses occur close enough together so that the concentration overpotential cannot recover to zero before the next pulse. This may be seen in the behavior of the polarization current in Figure 4.3d. Note that at the end of the pulse, the battery voltage does not recover to the OCV before the next pulse. This results in a drop in power at a given current along the train of pulses. The type of behavior illustrated in Figure 4.3c has implications for applications where short, high power pulses occurring in a train are required. Examples of such current pulses for a hybrid electric vehicle city drive cycle are shown in Figure 4.3e. Thus, hybrid vehicle applications require that the battery recovers rapidly from the short, but intense, current pulses associated with high power charge–discharge cycles. Otherwise, the effectiveness of the electrical power source will degrade over the drive cycle even if the SOC remains high. Improvement of the pulse power capability can be achieved by developing battery modules, where cells are linked in parallel or series and the battery model developed in Eqs. (4.1)–(4.5) can be used to examine this effect.

4.4.1
Parallel Connection

Consider the application of the FreedomCAR battery model to three cells linked in parallel in a battery module. In this case, the current from the three cells is

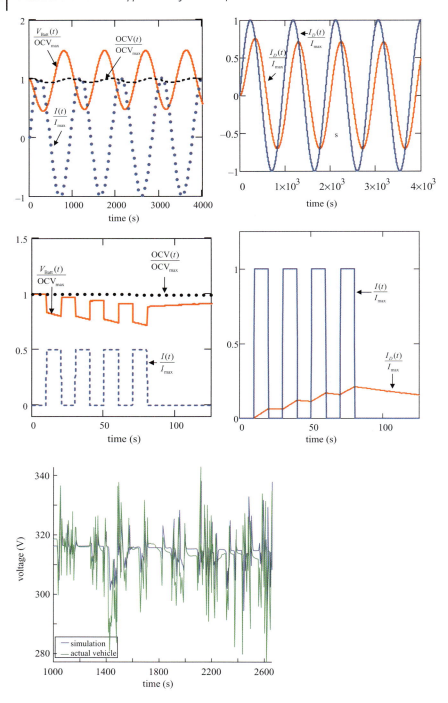

Figure 4.3 (a) Current–voltage response from the FreedomCAR equivalent circuit model letting $R_D = 1.5 \mid R_{ct} = 0.5 \mid C_D = 100\,F$, $C = 1500\,F$, $I(t) = \sin(t/150)$ A. The current, battery voltage, and open circuit voltage are normalized with respect to the maximum current and maximum OCV to give unitless quantities. Note the deviation of the measured battery voltage from the OCV value at a given current. Also note the phase lag between the peak of the current function and the trough of the voltage response. Both of these phenomena are primarily due to transport of charge in within the battery. (b) Demonstration of the polarization current, $I_D(t)$, as compared to the battery current, $I(t)$, from the Thevenin equivalent circuit presented in Figure 4.2. The phase difference observed here is responsible for the phase shift observed in (a). (c) A train of current pulses with voltage response simulated by the FreedomCAR battery model. The pulse train is shown to result in temporary battery voltage degradation even with minimal change in the OCV or SOC. This will degrade the power capability as the pulse train extends and this power degradation must be minimized for effective battery applications. (d) Demonstration of the time-dependent growth and decay of the polarization current, $I_D(t)$, within a battery current pulse train. The polarization current is proportional to the concentration overpotential. (e) Hybrid vehicle battery voltage as a function of time during a drive cycle. The large positive and negative deviations from the average voltage (approximately the OCV) is due to current pulses associated with rapid change from discharge to charge as driving conditions vary; taken from [16].

---

added but the voltage must be the same across all the three cells. If all the cells have an equivalent OCV and internal impedance, then the current need only be one-third that of the single cell battery examined in Figures 4.3a–d to achieve the same current pulse from the module. The effect of reducing the current on each individual cell of the pulse power response for the parallel connection of three batteries is shown in Figure 4.4. It can be seen that the reduced cell current leads to a smaller deviation for the three cell battery module as compared to the single cell battery; thus leading to a decrease in power degradation along the pulse train. Even though the voltage on the three cell module is the same as the voltage on each individual cell that is connected in parallel, the decrease in voltage loss due to the internal impedance leads to a higher power output as compared to a single cell battery.

4.4.2
Series Connections

As we have seen, high power applications utilizing batteries may require development of battery modules utilizing multiple electrochemical cells to meet power demands. Improvements in power capability achieved by connecting battery cells in series are similar to improvements obtained by a parallel battery connection. However, in the case of a series connection of batteries, the electrode reaction in each battery is coupled to the other batteries through the transfer of electrons. Consider, for example, the half-cell reactions for a Li-ion negative electrode during discharge:

$$LiC_6 \rightarrow Li^+ + e^- + C_6 \text{ Cell (2)} \tag{4.6}$$

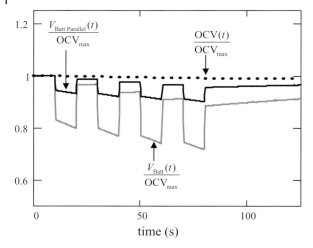

Figure 4.4 Equivalent circuit model of the pulse power capability of a single battery cell, V_{Batt}, compared to a parallel, three cell, battery module $V_{Batt\ Parallel}$ for the same magnitude of current pulse.

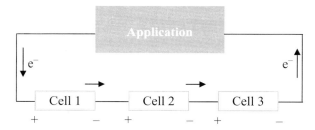

Figure 4.5 Schematic of coupled electron transfer of a battery module with three cells linked in series.

If this half-cell reaction occurs in the center cell (cell 2) in Figure 4.5, then the electron produced by the Li ionization is transferred to the positive electrode in cell 1 as these two electrodes are coupled by the attachment of the cell 2 negative electrode to the cell 1 positive electrode current collectors. This means that the cell 1 positive electrode reaction is coupled to the electron transfer from cell 2

$$CoO_2 + Li^+ + e^- (\text{from cell 2}) \rightarrow LiCoO_2 \text{ Cell (1)} \qquad (4.7)$$

The electron from the ionization process at the cell 1 negative electrode is then transferred to the positive electrode of cell 3 through the external application. And of course to complete the loop, the electron produced by the negative electrode ionization in cell 3 is transferred to the positive electrode in cell 2. The sequential electron transfer among the half-cell chemical reactions at each electrode means that the voltages for all the half-cells are added. The result is that the OCV of the three cell battery module is three times the OCV of an individual cell, assuming all the cells have equivalent properties. This situation of three cells linked in series may be modeled using the FreedomCAR equivalent circuit model, where the ideal voltage source has a magnitude increase of a factor of 3 over that of a single cell. However, since the cells are in series, then the resistive components of the equivalent circuits for each of the cells are also in series. Since these equivalent circuits obey all the rules of an electrical circuit, then these resistive components are also added. This means that the R_{ct} and the R_D are increased by a factor of 3 for the three cell series battery module for calculating the battery voltage in Eq. (4.3). On the other hand, only one-third of the current is required to achieve the same initial power as for a single cell battery at the initiation of the first current pulse. Therefore, unlike a single cell battery, a series connection of cells in a battery module has two competing influences; namely, the increase in resistance which reduces the power capability, and the increase in OCV which increases the power capability. The response of the equivalent circuit model for the battery module at one-third the current of a single cell battery (to get equivalent initial theoretical power) for the pulse train is compared to the single cell battery in Figure 4.6. It is again observed that the three cell module demonstrates an improvement in pulse power degradation as compared to the single cell battery at equivalent initial theoretical power.

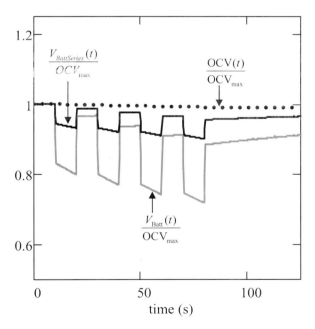

Figure 4.6 Equivalent circuit model of the battery voltage response to a train of current pulses. The deviation from the OCV results from the internal battery resistance to the transport of charged species.

It is interesting to see that the pulse power degradation for the series module (Figure 4.6) is the same as the parallel module at the same initial power output (Figure 4.4). Thus, for the same battery power, the battery module in series and the battery module in parallel reduce the pulse power degradation problem to the same extent. However, the series battery module current for this example is one-third that of the parallel battery module. As has been noted previously, a lower current higher voltage battery power solution is preferred to a higher current lower voltage solution even if the power for the two cases is equivalent. This concept can be illustrated through an electric motor application, as might be found in a hybrid electric vehicle. An electric motor has an electrical resistance associated with the wire windings responsible for the electromagnetic field driving shaft rotation. The current flowing through the windings within the electric motor has a power dissipation of

$$\text{Power Loss} = I(t)^2 R_{\text{windings}} \tag{4.8}$$

This means that the power output for the motor is related to the battery voltage, V_{Batt}, and battery current, $I(t)$, as

$$\text{Motor Power} = I(t)V_{\text{Batt}} - I(t)^2 R_{\text{windings}} \tag{4.9}$$

It is apparent from Eq. (4.9) that the current provided by the battery to the electric motor contributes to the motor power through both the product with the voltage and the product with the motor resistance. The motor power for the single cell battery, the three battery parallel module, and the three battery series module is compared in Figure 4.7 for the same theoretical power. The practical power response shown in Figure 4.7 not only indicates the power degradation due to the

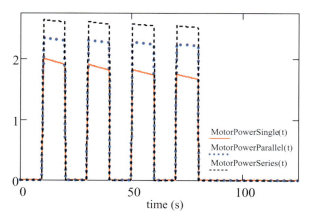

Figure 4.7 Power output of the electric motor for the same theoretical power output of the battery. The theoretical power at the battery is the product of the current and the OCV, but power dissipation occurs at both the battery internal charge transfer and the internal resistance of the motor, leading to the observed difference in behavior among the three configurations plotted above for the equivalent circuit model.

internal battery resistance described in Figures 4.4 and 4.6, but also shows the power loss from the internal motor resistance. To reiterate, this result demonstrates the advantage of a series battery configuration as compared to a parallel battery configuration as a solution to a particular high power application. The series solution provides a specific power at a lower current than the parallel solution, and thus the power loss during the application is lower.

4.4.3
Limitations and Safety Issues

There are of course limitations to series connection solutions for high power applications. For example, the DOE has placed a maximum of 400 V as a goal for hybrid electric vehicle applications. The reasoning behind the upper voltage limit is related to safety. Even though the series configuration is capable of providing a high power at a low current due to the high voltage, these high voltages are also capable of driving the current through very high resistances with the production of a large amount of heat. An additional limit to the series configuration is the probability that all the cells are not equivalent, and the possibility that a cell is "bad." Definitions of a bad cell would be that the resistance in that the cell is relatively large or that the capacity in that cell is relatively small. Using the three cell series module example for the case where an electrode contamination (reaction) layer impedes diffusion of Li-ions and charge transfer in one bad cell, the battery voltage for the module would have the form

$$V_{\text{Batt Series}}(t) = 3\text{OCV}(t) - [I(t)(2R_{ct} + R_{ct \text{ Bad Cell}})] - I_{D \text{ Series}}(t)(2R_D + R_{D \text{ Bad Cell}})$$

where the OCV(t) and C_D terms are equivalent in each cell. Clearly, the overpotential at a given current has increased for the module due to the presence of a single bad cell with increased resistance.

The performance of a battery module with cells in series may also be severely impacted by a bad cell, where the OCV(t) function of one cell decreases relatively rapidly with decreasing SOC as compared to the other cells in the series. The critical issue here is that when one cell in the series has a SOC of zero, then the electron coupling process across the cells described by Eqs. (4.6) and (4.7) is interrupted. That is, there is no raw material available for the half-cell electrode processes that were designed for the cell chemistry when the SOC is zero. The remaining cells in the series will still pump electrons (on discharge) into the positive electrode and pull electrons from the negative electrode, but the chemical or electrical response inside the "bad" cell will be problematic. One possible response to the electron transfer driven across a "bad" cell can be that the electrochemical breakdown of the electrolyte and other cell materials will occur within the bad cell. A second, or simultaneous, undesirable response that occurs in the bad cell due to the electron transfer driven across the cell is that the bad cell will behave as a capacitor of very low capacitance. That is, the lack of available electrochemically active material leads to a buildup of charge across the separator in the bad cell. This polarization would lead to a rapid drop in the overall battery module voltage due to a single bad cell. The destruction of cell materials in a bad cell that could occur if "unplanned" electrode processes are driven by the coupled charge transfer between batteries in series would have deleterious consequences. The electrolyte or electrode material breakdown may result in a gaseous by-product, causing an increased pressure within the cell. Furthermore, charging of such a damaged cell may prove to be impossible. To avoid the effects of a bad cell in a series configuration for high power applications, the series is coupled to a control system that monitors the SOC of each battery cell within the series. The control system should be capable of operational warnings and may be capable of shunting current around a bad cell, insuring safe and effective operation.

Li-ion battery safety for HEVs and other applications does remain a concern. The extensive recalls of laptop batteries in 2006 have indeed sharpened the consumer concerns related to Li-ion battery utilization. The major focus of Li-ion battery safety is associated with the energy density and the cell materials. The energy density of a Li-ion battery is approximately six times that of a lead acid battery and the sudden release of this energy due to a short circuit would generate significant heat [17]. The organic electrolyte used in the cell chemistry of current generation Li-ion batteries is flammable, and the heat pulse from a short circuit may result in ignition, particularly if the cell has been ruptured [18], as seen in Figure 4.8. In particular, the safety issues have been categorized into four areas [17]: overcharging, overheating, electrical short, and physical abuse.

1) The overcharging issue is related to the chemical instability of the fully delithiated (oxidized) positive electrode. The oxidized positive electrode will generally exist in a highly unstable crystal structure and the rapid transformation of this unstable structure to the equilibrium structure may produce enough heat to volatilize and ignite the cell components. Mitigation of this danger is carried out by smart charging systems, which monitor the voltage in each cell or parallel modules to insure that enough Li is left in the positive electrodes so that the crystal structure is partially stabilized. It is also considered that phosphate-based positive electrodes that have been delithiated by overcharging are more stable than the delithiated $LiCoO_2$. In particular, during charging,

Figure 4.8 One safety concern is the ignition of gases released from overcharged batteries that can produce flame clouds [18].

(a) LiCoO$_2$ (b) LiFePO$_4$

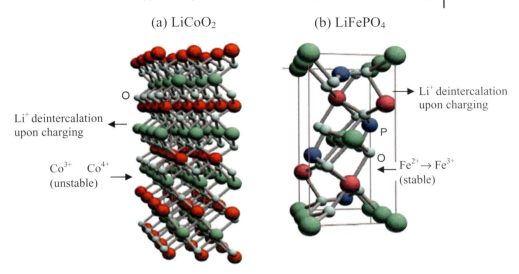

Figure 4.9 The cobalt-based and iron-based lattice of the cathode upon lithium deintercalation, during charging; taken from [19].

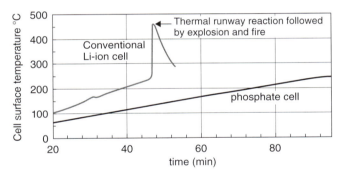

Figure 4.10 Comparing the safety of a conventional Li-ion cell with a phosphate cell; taken from [19].

cobalt-containing cathodes experience an oxidation from Co^{3+} to the unstable Co^{4+}, while phosphate-based cathodes undergo an oxidation from Fe^{2+} to the stable Fe^{3+} resulting in a safer, fault tolerant chemistry [1]; Figure 4.9. The stability of the cathode leads to less thermal runaway problems as demonstrated in Figure 4.10. The active materials are intrinsically stable and can be reduced to nanoscale sizes. This enhances the conductivity of the material due to the elevated total surface, adding to increased power without safety compromises [19]. Of course, as a consequence the operating voltage for a phosphate electrode is lower than that of an oxide electrode.

2) Overheating of a battery may occur if the heat produced due to the battery and device internal resistance is not efficiently dissipated. If the cell temperature increases over 130 °C, then the electrolyte reacts exothermically with Li in the negative electrode. This heat release is again a danger because of the volatilization of the flammable organic electrolyte. Mitigation of this danger during operation requires an effective cooling system and the possibility of a temperature-based battery control system that is capable of shutting down the battery system.

3) As noted above, short circuits are capable of causing rapid temperature increases in the battery and a possible thermal event. The use of circuit breakers to protect the battery system from an external short circuit is an effective means of mitigation. However, internal short circuits where a conductive material has breached the separator between the positive and negative electrode are quite problematic.

4) This is particularly true when mechanical abuse has caused both cell rupture and a short circuit. In this case, rapid heating, volatilization, and oxygen are all present and capable of producing a flame cloud. In a hybrid vehicle application, the battery system must have at least the same crash protection as that given to the vehicle gas tank. It is tempting to draw the analogy between the fire danger from gasoline and the fire danger from a high energy density battery. However, in the case of gasoline, the storage area is separated from the point of heat production. In the case of the high energy density battery, heat production occurs at the point where the "fuel" is stored, that is at the electrodes. Although safety concerns will continue into the future for high energy density battery applications, research progress in the development of Li-ion battery controls, electrodes, electrolytes, and separators [17] may lead to safety levels in hybrid vehicles comparable to gasoline.

The following chapters summarize the most promising candidate components (anodes, cathodes, electrolytes) for the next generation secondary Li batteries.

References

1 Nazri, G.-A., and Pistoia, G. (eds) (2009) *Lithium Batteries, Science and Technology*, Springer, Berlin.

2 Croce, F., D'Epifanio, A., Hassoun, J., Reale, P., and Scrosati, B. (2003) *J. Power Sources*, **119**, 399.

3 Motorola Cell Phone (2010) www.motorola.com (accessed January 2010).

4 Apple (2010) http://www.apple.com/macbookair/ (accessed January 2010).

5 Tesla Motor (2010) teslamotors.com (accessed January 2010).

6 Kintner-Meyer, M. (2007) Impacts Assessment of Plug in Hybrid Vehicles on Electric Utilities and Regional U.S. Power Grids, November 2007, Pacific Northwest National Laboratory.

7 Aquilina, O. (2006) *Images Paediatr. Cardiol.*, **27**, 17.

8 Greatbatch, W., Holmes, C.F., Takeuchi, E.S., and Ebel, S.J. (2006) *Pacing Clin. Electrophysiol.*, **19**, 1836.

9 National Institute of Health (2009) http://www.ninds.nih.gov/disorders/

deep_brain_stimulation/deep_brain_
stimulation.htm (accessed October
2009).

10 Gildenberg, P.L. (2005) *Stereotact. Funct.
Neurosurg.*, **83**, 71.

11 Food and Drug Administration (2009)
www.fda.gov (accessed October 2009).

12 Mayberg, H.S., Lozano, A.M., Voon, V.,
McNeely, H.E., Seminowicz, D.,
Hamani, C., Schwalb, J.M., and
Kennedy, S.H. (2005) *Neuron*, **45**, 651.

13 Velasco, F., Velasco, M., Velasco, A.L.,
Jimenez, F., Marquez, I., and Rise, M.
(1995) *Epilepsia*, **36**, 63.

14 Malone, D.A., Jr., and Pandya, M.M.
(2006) *Adv. Neurol.*, **99**, 241.

15 Schiff, N.D., Giacino, J.T., Kalmar, K.,
Victor, J.D., Baker, K., Gerber, M., Fritz,
B., Eisenberg, B., O'Connor, J.,
Kobylarz, E.J., Farris, S., Machado, A.,
McCagg, C., Plum, F., Fins, J.J., and
Rezai, A.R. (2007) *Nature*, **448**, 600.

16 Hunt, G., U.S. Department of Energy
(2003) FreedomCAR Test Manual, DOE/
ID 11069.

17 McDowall, J., Biensan, P., and Brous-
sely, M. (2007) *IEEE, Telecommunications
Energy Conference, 2007. INTELEC 2007.
29th International, October 4, 2007*, pp.
701–707.

18 Townsend, A., (2008) http://www.
popularmechanics.com/science/
research/4282985.html (accessed
October 2009).

19 A123 Systems (2009) www.A123systems.
com (accessed January 2009).

5
Li-Ion Cathodes: Materials Engineering Through Chemistry

Stephen A. Hackney

5.1
Energy Density and Thermodynamics

The positive electrode in the commercial Li-ion battery is based on a transition metal (M) oxide or a phosphate that has the composition $LiMO_2$ or $LiMPO_4$ in the discharged state. This chemistry is the basis for many of the high-voltage batteries developed in the last 20 years. The oxide and phosphate crystal structures act as a host for Li atoms in that the Li ions may be added and then removed in these crystal structures many times without causing significant or permanent changes in lattice symmetry. As discussed in Chapter 1, the capacity of a battery refers to the total amount of charge that the battery can deliver. There is a "theoretical capacity" corresponding to a thermodynamically reversible discharge (a nonmeasurable discharge rate), and actual battery capacities that depend on discharge rate. If a battery is discharged at a constant current, I, until the voltage falls to zero after a time, t, then the capacity, Q (in units of ampere-hours Ah), is given as

$$Q(Ah) = It$$

Consider that an electrochemically positive electrode material in a fully charged battery has the composition CoO_2 that will undergo complete discharge to the composition $LiCoO_2$. The theoretical capacity of an electrode with 10 g CoO_2 active material can be determined with some simple chemistry considerations. Given a molecular weight of 91 g/mol for CoO_2, the number of moles of monovalent Li ions that will be taken up by the CoO_2 may be determined by their ratio. Note that a mole of monovalent species has a charge of 9.649×10^4 C (which is the Faraday constant, F); where a coulomb (C) is the charge carried by a current of 1 A in 1 s. So C has base units of ampere second and F has base units of A s/mol. A single monovalent ion (an electron or Li^+) has a charge of 1.602×10^{-19} C. This charge multiplied by Avogadro's number gives the Faraday constant. So we have

$$\left(91\frac{gm}{mol}\right)^{-1} 10gm \; 9.65 \times 10^4 \frac{C}{mol} = 1.06 \times 10^4 \; C$$

High Energy Density Lithium Batteries. Edited by Katerina E. Aifantis, Stephen A. Hackney, and R. Vasant Kumar
© 2010 WILEY-VCH Verlag GmbH & Co. KGaA, Weinheim
ISBN: 978-3-527-32407-1

$$1.06 \times 10^4 \,C\frac{As}{C}\frac{1h}{3600s} = 2.94\,Ah$$

Although the capacity of the battery is an important consideration, it is only part of the equation for how much work a battery can do. Thermodynamics tells us that a reversible, incremental change in free energy, dG, due to a chemical reaction in the electrodes is related to maximum work, w_{max}, through the expression

$$-dG = \delta w_{max}$$

We can relate this work that is available from the chemical reaction in the battery to the electrical work associated with slowly transporting dn ions (the Li ions in this case) against a potential difference, V as

$$\delta w_{max} = zFV(n) = -dG(n) \tag{5.1}$$

where F is Faradays' constant and z is the valence of the ion. The voltage in Eq. (5.1) is also the voltage that would be observed if the external circuit were "open." When the circuit is open, no ion or electron current can flow and there is no "wasted" energy and the measured voltage is expected to correspond to the maximum possible voltage. This is known as the "open-circuit potential," or OCV. The capacity is thus a measure of the total amount of charge that will move through a battery, and the energy of a battery is the product of the amount of charge and the electrochemical potential that the charge is capable of moving against. Thus, the energy available from a particular battery is

$$\text{Energy} = \int_0^N zFV(n)\,dn$$

where N is the maximum charge available in the battery.

For a constant current (I) discharge, the rate at which the charge is passing through the potential is known and a change of variable is possible so that the energy is obtained by integrating the power for the discharge cycle of the battery occurring over a time, t, in hours. If the voltage, $V(t)$, is time dependent, then

$$\text{Wh} = I\int_0^t V(\tau)\,d\tau$$

where Wh refers to Watt-hours. So it is apparent that the Wh for a battery discharged at a constant current will depend on the magnitude of the voltage over the time required to complete the discharge and the magnitude of the product of current with time. The Li-ion battery works because the Li component forms a very stable compound with MO_2, while the Li in the negative electrode is comparatively unstable. In terms of chemical thermodynamics, this means that the Gibbs free energy for the formation of $LiMO_2$ or $Li\,MPO_4$ is a very large negative number compared to the free energy of formation for electroactive negative electrode in the discharged state. If we consider the reaction equilibrium between the components of LiC_6 (negative electrode) and MO_2 in a system in which the moles of LiC_6 and MO_2 are initially equal:

$$LiC_6 + MO_2 \leftrightarrow LiMO_2 + C_6 \tag{5.2}$$

then we expect reaction (5.2) to proceed spontaneously to the right. The Gibbs free energy change required for the reactants, LiC_6 and MO_2, to come to equilibrium with the product $LiMO_2$ is defined in terms of the free energy required to create MO_2 from pure M and O, LiC_6 from pure Li and C, and the free energy required to create $LiMO_2$ from pure Li, M, and O. The "products" are the components on the right hand side of Eq. (5.2), while the reactants are the components on the left hand side of Eq. (5.2). The free energy of formation of pure elements is defined to be zero:

$$\Delta G = G_{products} - G_{reactants} \tag{5.3}$$

This is the theoretical amount of energy available to do work from reaction (5.2). However, if we want this reaction to perform electrical work, then we have to arrange the materials in a configuration that forces electrons to flow through a circuit outside the system. Let us rewrite reaction (5.2) as the sum of two separate reactions, which involve Li ionization

$$LiC_6 \rightarrow Li^+ + e^- + C_6 \tag{5.4a}$$

$$Li^+ + e^- + MO_2 \rightarrow LiMO_2 \tag{5.4b}$$

In this case, reaction (5.4a) occurs on the negative electrode and reaction (5.4b) occurs on the positive electrode and are thus called the "half-cell reactions." If the two reacting species, LiC_6 and MO_2, are separated by a material that is both an Li-ion conductor and electric insulator inside the battery, but connected by an electrical system outside the battery, then we have a device in which the chemical reaction drives the motion of electrons and ions, and the voltage at which the charged species move is dependent on the free energy change for the reactions. That is, the electrons produced in reaction (5.4a) can be driven through an external device because they are being annihilated in reaction (5.4b). In terms of battery operation, reactions (5.2), and (5.4a,b) proceeding to the right would be a discharge process as the associated current external to the battery will provide energy for useful work. If the battery is "charged," then reactions (5.2) and (5.4a, b) proceed to the left and the electrons and Li ions must be forced from the positive electrode to the negative electrode by an external power supply (such as a battery recharger for a cell phone or a brake system generator in a hybrid vehicle).

When the electrochemical processes (5.4a,b), and associated charged particle transport occur very close to equilibrium (very slowly), they are said to occur "reversibly." Thermodynamically reversible processes allow the maximum possible work available from the battery to be recovered. In practice, we will never be able to attain this maximum amount of work when running a consumer electronics device or hybrid vehicle because the processes inside the cell are forced to occur rapidly and far from equilibrium, or "irreversibly," leading to a transformation of some of the available electrochemical energy into waste heat. This waste heat production leads to the reduction in the useful energy relative to the maximum work that can be accomplished. However, the consideration of the maximum possible useful energy available in a battery is nevertheless of value because it provides a benchmark to examine cell efficiency and a quantitative method to compare various battery materials.

The total amount of transported charge as a percentage of the maximum available charge, subtracted from a 100%, gives the "state of charge" (SOC) of the battery. Notice that the voltage and free energy are written as a function of the number of ions transported across the cell. This is the case for many modern battery chemistries where the transported ions form a "solution" or multiple compounds with the electrode material. It is also the cause of the OCV of a battery being a function of the SOC. An example of a voltage/time curve relating voltage and SOC from a commercial Li-ion battery being discharged in a cell phone is shown in Figure 5.1b. This battery has a $LiCoO_2$-positive electrode and a carbon-negative electrode. We can study the relationship between SOC and OCV in a battery by studying the negative and positive electrodes separately. For a Li-ion battery like the one being discussed here, this usually involves studying the discharge behavior of the carbon electrode and the discharge behavior of the oxide with Li as the counter electrode. The OCV/x curves for the oxide and carbon electrodes being discharged against an Li electrode are shown in Figure 5.1b. It is seen that the OCV for discharge against Li is a function of SOC for both electrodes. The fall off in OCV with SOC of the commercial battery can now be seen as a combination of behavior of both the anode and cathode as shown in Figure 5.1c.

It is apparent from Figures 5.1a–c that the voltage of the cell is dependent on the SOC, falling almost a full volt before the phone undergoes auto shut-off in Figure 5.1c. On the other hand, the figures demonstrate how materials chemistry and structure may be engineered to modify the energy density of the cell. The "sloping" voltage/capacity curve of a commercial $LiCoO_2$/carbon battery is considered to be a fundamental limitation of the pulse power capability required for transportation applications. The pulse power capability, P_c, on discharge is defined as

$$P_c = V_{min}(OCV - V_{min})R_{batt}^{-1}$$

where V_{min} is the minimum allowed voltage on the battery and R_{batt} is the internal resistance of the battery determined using a current pulse test. Thus, it is apparent that if the OCV decreases with SOC, the pulse discharge capability will be compromised at low SOC. As such, battery chemistries that provide a more planar voltage plateau in the discharge curve are also being developed. Examination of Figures 5.2 and 5.3 show that electrode materials such as $LiMn_2O_4$- and $LiFePO_4$-based materials may be chosen or designed, which provide voltage plateaus that are relatively flat, as opposed to the sloping profiles exhibited by $LiCoO_2$-based materials.

In order to explain the thermodynamic basis for the contrasting observations among Figures 5.1–5.3, we have to understand that the concentration of Li is not fixed within the host electrode materials but rather can change during the discharge. This gradual change in concentration of the reactant and the product during discharge effects the free energy change (and voltage) during each stage of

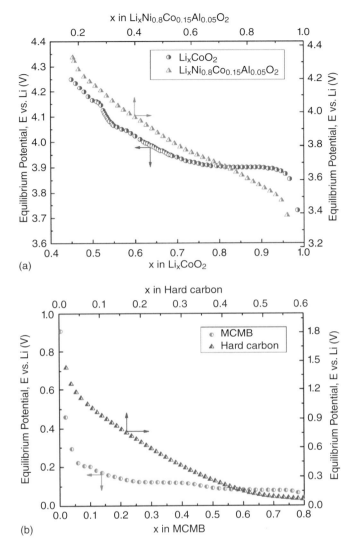

Figure 5.1 (a) OCV curve for oxide-based positive electrode materials as a function of Li fraction (X). From Ref. [1]. (b) OCV curve for carbon-based negative electrode materials as a function of Li fraction (X). From reference [1].

the discharge process. To treat voltage:capacity phenomena explicitly for these materials, the behavior of LiCoO₂ versus Li is first considered. In order to describe Figure 5.1a, consider the positive electrode, where the half-cell reaction is

$$CoO_2 + Li^+ + e^- \rightarrow LiCoO_2 \tag{5.5a}$$

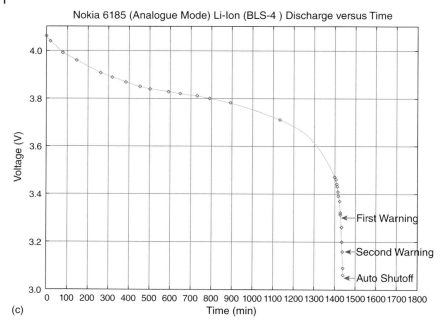

Figure 5.1 (c) Discharge curve for a commercial Li-ion battery in a cell phone (Nokia BLS-4 Battery). Taken from [2].

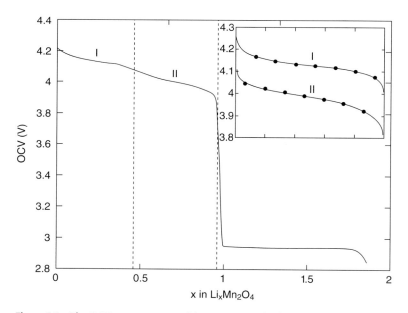

Figure 5.2 The OCV versus *x* curve of the Li/Li$_x$Mn$_2$O$_4$ half cell at the ambient temperature obtained by GITT. From reference [3].

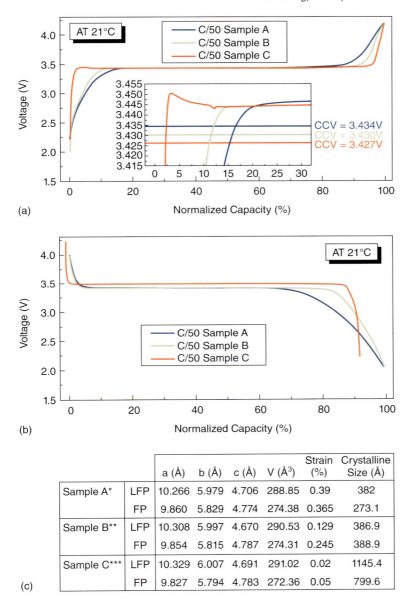

		a (Å)	b (Å)	c (Å)	V (Å³)	Strain (%)	Crystalline Size (Å)
Sample A*	LFP	10.266	5.979	4.706	288.85	0.39	382
	FP	9.860	5.829	4.774	274.38	0.365	273.1
Sample B**	LFP	10.308	5.997	4.670	290.53	0.129	386.9
	FP	9.854	5.815	4.787	274.31	0.245	388.9
Sample C***	LFP	10.329	6.007	4.691	291.02	0.02	1145.4
	FP	9.827	5.794	4.783	272.36	0.05	799.6

(c)

Figure 5.3 Galvanostatic charging (a) and discharging (b) at C/50 rate for LiFePO₄ shows the narrower composition range over which smaller size particles exhibit a constant cell voltage corresponding to two-phase coexistence. (c) XRD Reitveld refinement for the A, B, and C samples. From reference [4].

One way to write the free energy of the electrode is in terms of the mixing of the pure components, CoO_2 and $LiCoO_2$:

$$G(X) = (1-X)G_{CoO_2} + XG_{LiCoO_2} + [RT[(1-X)\ln(1-X) + X\ln(X)]] + RT\ln(\gamma(X))$$

(5.5b)

where γ is the activity coefficient describing the chemical interaction between the pure components. The three component-specific free energy terms in Eq. (5.5b) are written in terms of the "pure" components Eq. (5.6a), a term associated with the entropy of mixing of those components Eq. (5.6b), and a term describing the chemical interaction of the pure components with one another (attraction, repulsion, etc.) in Eq. (5.6c):

$$(1-X)G_{CoO_2} + XG_{LiCoO_2}$$

(5.6a)

$$[RT[(1-X)\ln(1-X) + X\ln(X)]]$$

(5.6b)

$$RT\ln(\gamma(X))$$

(5.6c)

The activity coefficient is an X-dependent term that is a measure of how the Li component interacts with the surrounding atoms in the crystal structure of the electrode materials as the Li concentration (X) changes. The activity coefficient can be a constant $\gamma = 1$ when the chemical interaction is negligible, and in this case, the mixture of components is said to be "ideal." Otherwise, it is a "nonideal" mixture.

Now we need to relate Eqs. (5.4)–(5.6) back to the cell voltage. Dividing both sides of Eq. (5.4) by the maximum possible number of Li ions that may be transferred across the separator, N, gives

$$zFV(n)dnN^{-1} = -dGN^{-1}$$

(5.7a)

$$zFV(X) = \frac{-dG}{dX}$$

(5.7b)

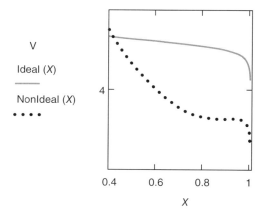

V

Ideal (X)

———

NonIdeal (X)

· · · ·

Figure 5.4 Comparison of the voltage-X behavior for an ideal and nonideal solution.

Substituting Eq. (5.5b) into Eq. (5.7b) determines the composition-dependent cell voltage, $V(X)$, in terms of thermodynamic properties

$$V(X) = -(zF)^{-1} \cdot \left[\Delta G_{\text{LiCoO}_2 - \text{CoO}_2} + \left[RT \left(\ln \left(\frac{X}{1-X} \right) + \frac{d}{dX} \ln(\gamma(X)) \right) \right] \right] \qquad (5.7c)$$

Equation (5.7c) is fit to the data in Eq. (5.3b) as shown in Figure 5.4 (dotted curve) for the case where

$$\text{OCV}(X) = \frac{1}{100000} \left(-2600 \ln \left(\frac{-X}{-1+X} \right) + 207000 X^2 - 344655 X + 508054 \right) V$$

This form of the activity coefficient is clearly describing a nonideal mixture. The voltage behavior for an ideal mixture is shown in Figure 5.4 (solid curve) for comparison.

5.2
Materials Chemistry and Engineering of Voltage Plateau

This sort of treatment of LiCoO_2 leading up to Eq. (5.7c) assumes that there is a continuous variation in free energy as x changes. This means that there are no discontinuous changes in crystal symmetry, or density or heat capacity with x. As such, the analysis gives the Li_xCoO_2 system the properties of a nonideal solid solution between CoO_2 and LiCoO_2. This assumption is the same thing as saying that there is no "phase transition" in Li_xCoO_2 as x changes. This assumption is likely to be false as there are reports of multiple changes in crystal symmetry as x changes. However, from a practical standpoint, the voltage behaves as a nonideal solid solution because of the sloping profile. This is due to the fact that the multiple phase transitions are close together in the "x" space, and thus the properties have the appearance of a continuous change. On the other hand, the plateaus of voltage in Figures 5.2 and 5.3 suggest a significant discontinuity in materials properties (changes in phase). To understand the reasons behind this phenomena of voltage plateau and the correlation with phase change, Ramana *et al.* [5] have used the Gibbs phase rule as a point of discussion, "$f = c - p + n$, with n being the number of the intensive variables necessary to describe the system, p the number of phases and c the number of components. In electrochemical studies, the intensive quantities are only temperature and pressure, but they are kept constant, so that $n = 0$. Here, we deal with a binary system ($c = 2$). ... If only one phase exists in a particle, $p = 1$ and $f = 1$ so that the potential is a degree of freedom and varies with the lithium concentration. On another hand, if the particles contain the two phases, $p = 2$, so that $f = 0$, in which case no intensive variable (e.g., potential) can change." In other words, planar voltage plateaus will exist when two phases coexist in a binary (two component) electrode material. In terms of a quantitative thermodynamic model, the free energy function for two phases, α and β, in equilibrium at fixed compositions, X_α and X_β, is given by

$$G(x) = G_\alpha \frac{(|x - X_\beta|)}{|X_\alpha - X_\beta|} + G_\beta \frac{(|x - X_\alpha|)}{|X_\alpha - X_\beta|} \tag{5.8a}$$

The derivative of $G(x)$ with respect to X that will determine the voltage via Eq. (5.7b) is then

$$\frac{d}{dx} G(x) = G_\alpha \frac{1}{|X_\alpha - X_\beta|} + G_\beta \frac{1}{|X_\alpha - X_\beta|} \tag{5.8b}$$

Since all the terms are constant on the right hand side of Eq. (5.8b), then the potential voltage will be constant as x varies, thus leading to the technologically preferred voltage plateau. The phosphate-based positive electrode material is particularly interesting because of the very long and very flat voltage profile occurring above 3 V versus Li. Several investigations have attributed this property to the presence of a miscibility gap in the phase diagram (Figure 5.5) by which the end members, FePO$_4$ (FP) and LiFePO$_4$ (LFP), tend to form clusters as opposed to randomly mixing.

When compositions of the electrode material move into the miscibility gap during charge or discharge, there is a tendency for the uniform composition to decompose spontaneously into composition fluctuations with the extremes approaching the solubility limits (Figure 5.5a). This has been studied by several investigations of the lithiation of iron phosphate. Chen et al. [6] investigated the chemistry of LiFePO$_4$ before and after chemical delithiation using TEM. At the composition of Li$_{0.5}$FePO$_4$, the authors observed stripes of alternating contrast parallel to (100) planes. These stripes of contrast were correlated to alternating regions of compositions of approximately LiFePO$_4$ and FePO$_4$ by imaging the variation in lattice spacing between the adjacent regions of dark and light contrast and relating lattice spacing with composition. The conclusions of Chen et al. [6] were supported by the work of Laffont et al. [7] using high-resolution electron energy loss, who also concluded that zones of alternating FePO$_4$ and LiFePO$_4$ composition are present in Li$_x$FePO$_4$ particles. When this type of composition variation arises by spontaneous decomposition of a uniform solution, it is known as spinodal decomposition. In this case, spinodal decomposition originates from the atomic scale interactions among LFP and FP components. It can be specified that a particular decrease in energy of the two component system may occur when the components are brought together into close proximity from being infinitely far apart. These negative energies may be termed $E_{LFP:LFP}$, $E_{LFP:FP}$, and $E_{FP:FP}$ for the permutations of the various component combinations. Using these definitions of energy of interaction, the expression for a system of composition Li$_x$FePO$_4$ equivalent to Eq. (5.6c) becomes

$$RT\ln(\gamma_{LFP}) = zA\left[E_{LFP:LP} - \frac{1}{2}(E_{LFP:LFP} + E_{FP:FP}) \right](1 - X)^2 \tag{5.9a}$$

and

$$RT\ln(\gamma_{FP}) = zA\left[E_{LFP:LP} - \frac{1}{2}(E_{LFP:LFP} + E_{FP:FP}) \right](X)^2 \tag{5.9b}$$

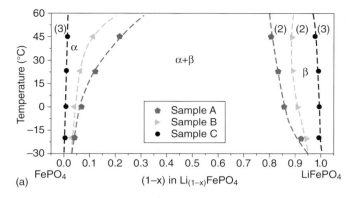

(a)

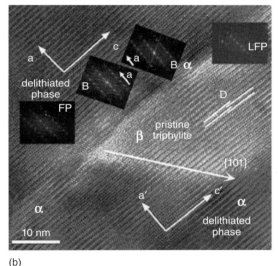

(b)

Figure 5.5 (a) Experimentally determined phase diagram for nanoscale lithium iron phosphate as a function of particle size. It is observed that the miscibility gap contracts systematically and solid solution limits increase with decreasing particle size and increasing temperature. From reference [4]. (b) HRTEM image of the 0.5 (LiFePO$_4$ + FePO$_4$) sample showing a LiFePO$_4$ cluster in the ac-plane. Inserts are the fast Fourier Transform of the selected regions, namely LiFePO$_4$ (LFP) domain, FePO$_4$ (FP) domain, and across the boundary (B). The splitting of spots of diffraction maxima along the a-direction due to a compositional difference along with lattice parameters (see text) is evident in the FFTs represented as "B" (for boundary). From reference [6].

where z is the coordination number and A is Avogadro's number. According to the Gibbs phase rule, a planar voltage profile is expected when an FP-rich phase can be in equilibrium with an LFP phase over a range of compositions (values of X). This is expected when the battery electrode chemistry allows

$$\left[E_{LFP:LP} - \frac{1}{2}(E_{LFP:LFP} + E_{FP:FP}) \right](X)^2 > 0 \qquad (5.10)$$

Meaning that as X increases from zero, a point will be reached where the LFP and FP components will begin to segregate into domains in order to lower the energy of the system. These segregated regions are in equilibrium with one another over a range of X, delivering the requirement for the planar voltage profile. The complicating factor is that the spacing between the components is a function of X, and therefore there will be a local distortion of the positions of the components if the crystals are to maintain structural compatibility. That is, unless the particles crack or develop other discontinuities in the lattice such as dislocations, there must be some compromise between the actual spacing of the components and the spacing of the components that gives the lowest possible energy. When the crystal lattice is continuous, the actual spacing of the components is then correlated to a materials "stress" in which the actual spacing of the components is different than the "stress-free" spacing of the components. As the spinodal decomposition process gives rise to spatial variations in composition, the stress-free lattice parameter is also expected to have correlated spatial variations. However, the variations in lattice parameter are constrained by the ionic bonding of the ions in the crystal. That is, the ions have a composition-dependent equilibrium spacing that gives the minimum free energy at zero stress, but the spatial variation in composition leads to a distortion away from the stress-free equilibrium positions. This behavior is reflected in the X-ray diffraction data presented in Figure 5.3c. The complicating factor here that the free energy and thus the voltage are functions of the stress. Continuum elasticity is a technique of analysis that provides a way to interpret the data presented in Figure 5.3c and to draw some conclusions about the role the decomposition process plays in the cell voltage profile in Figure 5.3b. Consider that the composition undergoes a periodic variation in one spatial dimension (say in the x direction) in a cathode particle (Figure 5.6) with the corresponding stress-free lattice parameter variations

$$C = \varepsilon \sin\left(\pi n \frac{x}{L}\right) + C_O \qquad (5.11)$$

where n is an odd integer and L is the half length of the crystal along the x and y directions.

Equation (5.11) will define the body force and body force potential such that

$$V = E\alpha\varepsilon \sin\left(\pi n \frac{x}{L}\right)$$

The solution to the equilibrium elasticity equation with body force potential is reviewed in Chapter 8 (Eq. (8.73) and Eq. (8.74)). The solution to this equilibrium equation must be obtained such that it is consistent with the stress-free surface boundary conditions. If no external forces are applied to the crystal surface, then

$$\sigma_{xx}(|x| = L) = 0$$

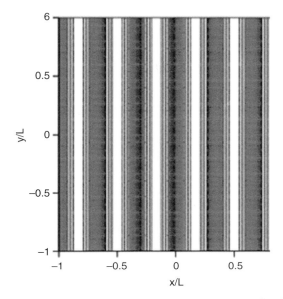

Figure 5.6 Periodic variation in composition (gray scale) along the x direction in a crystal having a half thickness of L.

$$\sigma_{yy}(|y| = L) = 0$$

$$\sigma_{xy}(|y| = L) = 0$$

Given these boundary conditions, it is expected that the general solution of Eqs. (8.72) and (8.73) of Chapter 8 to have the form

$$\varphi = \sin(\lambda x)(A\cosh(\lambda y) + B + Dy\sinh(\lambda y))$$

where

$$\lambda = \frac{n\pi}{L}$$

with n being an odd integer. The stress is then defined through the derivatives of the Airy stress function as

$$\sigma_{xx} = [D(2\lambda\cosh(\lambda y) + \lambda^2 y\sinh(\lambda y)) + A\lambda^2\cosh(\lambda y) + E\alpha\delta]\sin(\lambda x)$$

$$\sigma_{yy} = -\sin(\lambda x)\lambda^2(A\cosh(\lambda y) + B + Dy\sinh(\lambda y)) + E\alpha\delta\sin(\lambda x)$$

$$\sigma_{xy} = [\sin(\lambda x)\lambda(A\sinh(\lambda y)\lambda + B\sinh(\lambda y) + By\cosh(\lambda y)\lambda)]$$

Based on the form of the body force potential

$$B = E\alpha\varepsilon(1 - v)\lambda^{-2}$$

Application of the stress-free surface boundary conditions then gives

$$A = E\alpha\delta\upsilon\frac{\sinh(\lambda L) + L\cosh(\lambda L)\lambda}{\lambda^2(\cosh(\lambda L)\sinh(\lambda L) + \lambda L)}$$

$$D = -\sinh(\lambda L)E\alpha\delta\frac{\upsilon}{\lambda(\cosh(\lambda L)\sinh(\lambda L) + \lambda L)}$$

To consider the result, let

$$k = n\pi$$

where n is the number of periods of composition variation within a single particle. Based on the results of Cahn [8–10], the value of k will scale with crystal size because the minimum wavelength of the composition variation is thermodynamically limited by the gradient energy. The primary stress component is along the y direction and is given by

$$\sigma_{yy} = -\sin\left(\frac{k}{L}x\right)A\cosh\left(\frac{k}{L}y\right)\left(\frac{k}{L}\right)^2 + \sin\left(\frac{k}{L}x\right)E\alpha\varepsilon\upsilon - \sin\left(\frac{k}{L}x\right)Dy\sinh\left(\frac{k}{L}y\right)\left(\frac{k}{L}\right)^2$$

where A and D now have the form

$$A = E\alpha\varepsilon\upsilon\frac{\sinh(k) + k\cosh(k)}{\left(\frac{k}{L}\right)^2(\cosh(k)\sinh(k) + k)}$$

$$D = -\sinh(k)E\alpha\varepsilon\frac{\upsilon}{\frac{k}{L}(\cosh(k)\sinh(k) + k)}$$

Thus examining the maximum primary stress along the center line of the crystal ($y = 0$) as a fraction of the maximum possible stress (the lattice parameter is uniform throughout the composition variation) using k as a variable parameter leads to the result illustrated in Figure 5.7.

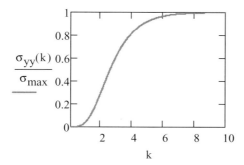

Figure 5.7 Correlation between primary stress component at the crystal center ($y = 0$) and $k = n\pi$.

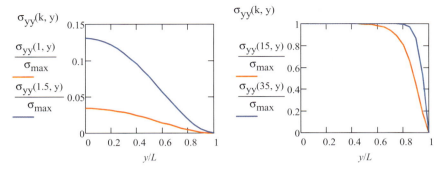

Figure 5.8 Distribution of principle stress magnitude across the crystal as a function of k.

The drop in stress at the center of the crystal as a function of crystal size is interpreted as the relaxation effect of the free surface. Such an interpretation is supported by examining the variation in stress across the crystal as a function of the y coordinate, where the stress reduction is apparent in Figures 5.8a and b as k decreases and the state of stress becomes much more uniform as k increases. The phenomena of particle size stress relief of internal body forces observed in the results examined in Figures 5.3 and 5.5 may now be considered as a materials engineering design method to influence the voltage behavior. Figure 5.3c shows a table in which the strain, as measured by XRD, increases as the crystal size becomes quite small. The XRD measures the variation in lattice parameter from the average lattice parameter, which we can see as a function of the surface relaxation effect in Figure 5.8. That is, a spatial variation in stress across the crystal shown in Figure 5.8 for small crystal size (small k) is proportional to the variation in strain across the crystal for small crystal size as seen in Figure 5.3c.

The role of the body stress and strain on the voltage profile is developed using the idea that the change in free energy per Li atom is changed by adding the amount of strain energy per Li atom to the stress-free chemical potential of Li or, equivalently, adding the volumetric strain energy density due to Li concentration variations to the stress-free volumetric free energy density. It has been shown for a crystal of infinite extent that the maximum volumetric strain energy density is

$$\frac{2E\alpha^2}{1-\upsilon}\Delta X^2$$

where ΔX is the maximum difference in Li mole fraction. The effect of strain on spinodal decomposition in solid materials has been considered for some time in the materials science literature. For spinodal decomposition to occur in a crystal of infinite extent

$$\frac{d^2}{dX^2}G_\upsilon + \frac{2E\alpha^2}{1-\upsilon} + 2\kappa\left(\frac{k}{L}\right)^2 < 0 \tag{5.12}$$

where G_v is the volume free energy and κ is the gradient energy term. The results above have shown that Eq. (5.12) must be modified when a finite-size single crystal particle is considered. For spinodal decomposition to occur at the center of the crystal, the critical condition becomes

$$\frac{d^2}{dX^2}G_v + 2\frac{E}{1-\upsilon}\alpha^2\left[-1\frac{\sinh(k)+1\cosh(k)k}{1(\cosh(k)\sinh(k)+k)}+1\right]^2 + 2\kappa\left(\frac{k}{L}\right)^2 < 0 \qquad (5.13)$$

where the second derivative of the G_v term is the only negative quantity. Taking $k = \pi$, κ = constant and considering the second derivative of G_v as a function of X allows the examination of the onset of spinodal decomposition as a function of SOC and crystal size. That is, rearrangement of Eq. (5.13) shows that spinodal decomposition cannot occur unless the crystal size exceeds the critical value L_c given by

$$L_c = \left[\left[(1-\upsilon)-4935\kappa\left[\frac{d^2}{dX^2}G_v(1-\upsilon)250+211E\alpha^2\right]^{-1}\right]^{\frac{1}{2}}\right]$$

For very small crystal size, the magnitude of the free energy second derivative must exceed a specific value as X increases before spinodal decomposition may initiate, thus limiting the extent of the planar voltage plateau as compared to a crystal of infinite extent. The result exemplified in Eq. (5.13) then rationalizes the observations in Figures 5.3b and 5.5a in that there is a critical crystal size for the onset of spinodal. Moreover, as the stress-free surface condition causes a stress relaxation within the crystal, the magnitude of volume-free energy change per Li intercalation is increased, leading to an increased voltage as seen in Figure 5.3a.

There is a further physically significant phenomena exposed by the XRD results in Figure 5.3c in that for larger crystal sizes, it is apparent that the average lattice parameter difference between the FP and LFP domains is greater at the larger particle size. One possible explanation is the formation of dislocation structures to relieve the elastic strain due to spatial variations in Li concentration. The presence of such dislocations has been shown by TEM by Gabrisch *et al.* [11] for large crystals (Figure 5.9a). However, it is known that the dislocation/surface interaction in a small crystal, known as an "image force," reduces the propensity for stress relief by dislocation motion. This suggests that the ability to form stress-relieving dislocations is hindered in small crystals, relative to large crystals and that the elastic strain associated with spatial variations in stress-free lattice parameter will be greater in the small crystal. On the other hand, the elastic stress in the small crystal is expected to be reduced relative to large crystals because of the stress-free boundary conditions. These competing methods of stress relief provide an interesting mechanics of materials problem that has yet to be resolved or even rigorously stated. In addition, the ability of dislocations to multiply and glide in response to the body forces in LP/LFP composite structures may affect the fracture toughness of the phosphate crystals. The image force from free surfaces may then embrittle small particles or the surface region of large particles, leading to particle degradation by cracking as shown in Figure 5.9b.

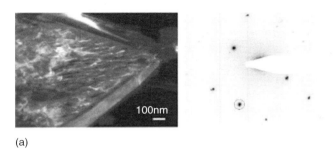

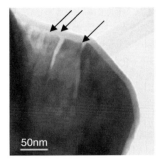
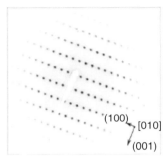

(100) [010]
(001)

(b)

Figure 5.9 (a) Dark-field TEM image and associated diffraction pattern showing the presence of dislocations (white lines) in particles retrieved from a chemically delithiated $Li_{0.5}FePO_4$ particle. (b) Surface fractures (indicated by arrow) on an electrochemically cycled $LiFePO_4$ crystal. From reference [11].

5.3
Multitransition Metal Oxide Engineering for Capacity and Stability

Thackeray *et al.* [12] have developed the approach that the properties of oxide cathode materials may be engineered using combinations of crystal structures and crystal chemistries. This type of approach is embodied in the work on structurally integrated $\chi Li_2MnO_3:(1-\chi)\cdot LiMO_2$ materials, where M can be either a single element, such as Mn, or a combination of elements such as $Mn_{0.5}Ni_{0.5}$. As may be discerned from the chemical formula, these materials exhibit more than one Li per metal cation, a ratio which could be very favorable for achieving higher specific capacity than oxides cycled over the composition ranges $MnO_2 \leftrightarrow LiMn_2O_4$; $CoO_2 \leftrightarrow LiCoO_2$ and $FP \leftrightarrow LiFePO_4$. Unfortunately, the oxidation state of Mn in Li_2MnO_3 is already at +4 and removal of Li from this structure without other changes in composition will lead to untenable Mn oxidation states. The component $LiMO_2$ for its part belongs to the family of $LiCoO_2$ and $LiNiO_2$, which are known to exhibit excellent electrochemical cycling characteristics, albeit at a lower theoretical capacity than Li_2MnO_3. The impetus to replace the Co in the positive electrode has led to efforts to create electrochemically active "layered" $LiMO_2$ structures (see Figure

(a) Li_2MnO_3 (b) $LiMnO_2$

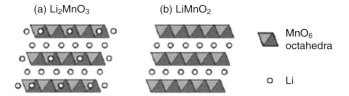

MnO₆ octahedra

Li

Figure 5.10 Layered structures of (a) Li_2MnO_3, (b) $LiMO_2$ (M = Co, Ni, Mn, or combinations of element). From reference [12].

5.10) where M is rich in Mn and Ni, but dilute in Co. The problem with this chemistry is that the delithiated layered structure, MO_2, is not thermodynamically stable.

The basic idea explored by Thackeray *et al.* [12] is that a composite material of the Li_2MnO_3 crystal structure and chemistry with that of the $LiMO_2$ component may combine the higher theoretical specific capacity of Li_2MnO_3 with the attractive electrochemical reversibility of the $LiMO_2$. The structure/chemistry of the two components is shown in Figure 5.10a. Although the crystal space group is different for each component, the configuration of the oxygen anions is quite similar between the two. According to [12], "The compatibility of the close-packed layers allows the integration of a Li_2MnO_3 component with a $LiMO_2$ component at the atomic level, if one allows for some disorder between the Mn and M cations. ..." Superimposing the charge–discharge cycle of the crystal structure nanocomposite on a Li_2MnO_3:$LiMO_2$:MO_2 ternary composition diagram illustrates some of the electrochemical properties of this system. It is observed that on the first charge there is an unexpected capacity in the Li_2MnO_3 0.7 $LiMn_{0.5}Ni_{0.5}O_2$ composite above 4.4 V versus Li, leading to a specific capacity of 285 mAh/g on charge. However, this high-voltage component of the capacity on charge is irreversible, as the first discharge shows a sloping voltage profile initiating at ~4.4 V with a capacity of 180 mAh/g. The irreversibility of the capacity in this material is associated with the removal of O anions from the crystal structure with the Li cations from Li_2MnO_3 component in order to maintain electrical neutrality in the crystal. This removal of oxygen is evidently thermodynamically favorable as compared to the Mn oxidation state increasing above 4+. Although the irreversibility of the capacity above 4.4 V means that this "excess" capacity will not directly contribute to the energy density of the system, there can be an indirect contribution through the mitigation of capacity loss through Li-rich SEI formation and other Li-scavenging processes in the battery. Moreover, if the charge–discharge cycle is limited to voltages below 4.4 V such that the Li in the Li_2MnO_3 component is not removed, then there is experimental evidence [13] and theoretical evidence that the extra Li will interact with the $LiMO_2$:MO_2 component mixture to "stabilize" the crystal structure. The crystal stabilization mechanism of the Li_2MnO_3 component on the delithiated $LiMO_2$ component involves the migration of Li ions from the Li-rich regions of the electrochemically inactive Li_2MnO_3 to the Li-poor region of the delithiated $LiMO_2$. This migration adds Li back into the unstable MO_2 layered crystal structure; "thereby providing the additional binding energy necessary to maintain structural stability,"; forestalling the onset of symmetry change and associated crystal distortion. However, as previously mentioned these mobile Li cations originating from the Li_2MnO_3 cannot leave the crystal during charge unless the charge voltage exceeds 4.4 V, resulting in a simultaneous removal of O anions to maintain electrical neutrality.

The presence of the stable Li in the vicinity of the Li_2MnO_3 clusters evidently provides an additional benefit relative to the dynamics of Li migration during charge/discharge processes. It has been proposed that the spreading of Li from the Li_2MnO_3 domains into the partially delithiated MO_2 domains results in Li occupying both octahedral and tetrahedral sites. According to [12], "The co-existence of tetrahedral and octahedral lithium in the lithium-depleted layers gives the electrode structure two-dimensional, quasispinel-like features, reminiscent of the three-dimensional tetrahedral (8a)–octahedral (16c) interstitial network of the Mn_2O_4 spinel framework ..., suggesting that the lithium-depleted layers of $xLi_2MnO_3(1 - x) Li_{1-2y}MO_2$ electrodes provide an energetically favorable interstitial space for lithium, thereby ensuring fast reaction kinetics. Indeed, there have been reports that a 10–15% excess lithium in $Li_{1+x}M_{1-2x}O_2$ electrodes enhances their rate capability consistent with our hypothesis." It should also be noted that the presence of this stable Li in the vicinity of the Li_2MnO_3 domains guarantees a mixture of +3/+4 Mn valence. This type of mixed valence has been shown via molecular dynamics simulation [14] to produce a high-frequency hopping of electrons between the Mn ions of differing valence. This oscillation of negative charge between adjacent sites is then predicted to cause a transient distortion in the oxygen anion sublattice and easing the migration of the mobile Li. According to these authors [14], "... Li diffusion occurred purely as a consequence of the lattice dynamics. It arose only when the spinel lattice was undergoing significant local distortion, such as during the initial 30 ps of structural relaxation, or following every reshuffle of the MnIII/MnIV valences in model II. This correlation supports a model in which Li migration is induced via a mechanism involving Mn, e.g., electron hopping and is mediated by resultant distortion modes of the mutually coordinating O atoms." It should be noted that this molecular dynamics simulation result is supported by observations that electrical transport properties in $LiMn_2O_4$ are dominated by nonadiabatic small polaron hopping [15]. It would thus appear that the migration of structural Li from Li_2MnO_3 domains into delithiated sections of the crystal will result in regions of high Li diffusivity in addition to providing a stabilization of the crystal structure.

The materials engineering possibilities of multicomponent oxides for Li battery applications have been recently amplified in the work of Wagemaker [16]. These authors have investigated the behavior of the $Li_xMg_{0.1}Ni_{0.4}Mn_{1.5}O_4$ spinel (P4332) chemically and electrochemically lithiated in the range $1 < x < 2.25$. These materials are interesting because the $LiMg Ni_{0.5} Mn_{1.5}O_4$ spinels exhibit a high voltage (4.7 V vs Li) and there is an ordering of Ni and Mn ions in the spinel octahedral sites, which reduces the space group symmetry from $Fd3m$ for random Ni and Mn octahedral site occupation. These properties and behavior are similar to that

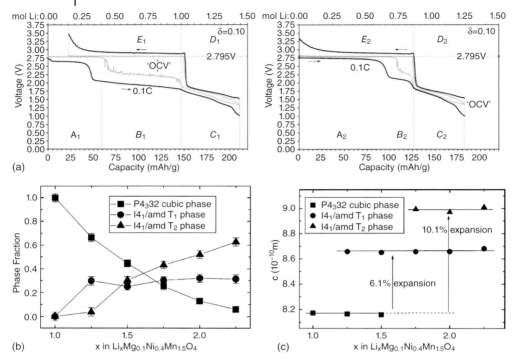

Figure 5.11 (a) Electrochemical insertion of lithium into $LiMg_{0.1}Ni_{0.4}Mn_{1.5}O_4$ between 3.5 and 1.0 V versus Li/Li^+: constant-current (black) and OCV (gray) curves for the first cycle (left) and the second cycle (right). (b) Phase fractions from Rietveld refinement as a function of the overall lithium content in the chemically lithiated $LiMg_{0.1}Ni_{0.4}Mn_{1.5}O_4$ samples. (c) Lattice expansion of the c axis of the two tetragonal phases relative to the original cubic phase as a function of the lithium content. The size of the symbols represents the size of the error bars.

found for the $LiNi_{0.5}Mn_{1.5}O_4$ spinel, so it is not really clear from the studies to date what the advantage or influence of the addition of Mg ions might be. Nevertheless, the studies of this material at lithium content above Li/(polyvalent metal) > 1 provide new insight into materials design possibilities. Wagemaker *et al.* find that there is a significant change in the OCV and discharge curve between the first and second cycle, as seen in Figure 5.11.

Detailed structural characterization of the material using Rietveld refinement of X-ray and neutron diffraction data indicates the formation of two new tetragonal phases (T_1 and T_2) simultaneous with the migration of Ni and Mn cations to form a "clustered" structure. The new-short range order of Ni and Mn destroys the ordered Ni:Mn structure present in the initial material. On the basis of the structural information, the authors of [16] propose the following scheme for an explanation of the observed cycling behavior. "The appearance of the T1 phase in the range $x = 1-1.25$ (5.11) suggests that the first potential plateau at about 2.7 V

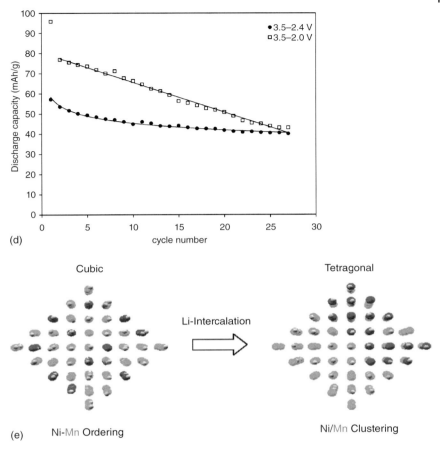

(d)

(e)   Ni-Mn Ordering   Ni/Mn Clustering

Figure 5.11 (d) Cyclability of LiMg$_{0.1}$Ni$_{0.4}$Mn$_{1.5}$O$_4$ versus Li/Li$^+$ with different lower cutoff voltages at a 0.1 C rate (solid circles, 3.5–2.4 V; open squares, 3.5–2.0 V). (e) Schematic of simultaneous cation clustering and symmetry change associated with increasing Li content. From reference [16].

versus Li/Li$^+$ is related to the coexistence of the initial cubic phase with this T1 phase. During the intercalation of lithium and the phase transition, the migration of the Mn and Ni ions destroys the initial Ni–Mn ordering. Further intercalation of lithium results in the formation of the T2 phase (Figure 5.11b), which has a relatively high Ni : Mn ratio compared to the initial phase. The formation of the Ni-rich T2 phase is consistently accompanied by a decrease in the amount of Ni in the T1 phase. Although intercalation in the range $x = 1.25$–2.25 predominantly results in the formation of the T2 phase (the T1 phase fraction grows only marginally), the T1 phase progressively takes up more lithium. ... Therefore, step B at about 2 V versus Li/Li$^+$ is not only associated with the formation of the T2 phase, but also reflects a mixed potential of lithium in both tetragonal phases. The fact that this plateau is not flat is ascribed to the variation of the lithium concentration in the T1 phase and the migration of Ni and Mn, which changes the Ni : Mn ratio in the T1 phase due to the formation of the Ni-rich T2 phase. The decrease of the average Ni : Mn ratio in the T1 phase, which means an increase of reducible Mn^{4+}, results in the accommodation of additional lithium in the T1 phase. At the beginning of the second cycle, the tetragonal phases are converted back into a cubic phase, with the difference that Ni-rich and Ni-poor domains have been formed in the first cycle. Subsequent lithium insertion again leads to the formation of the T1 phase, although this time is not hindered by the formation of the Ni-rich and Ni-poor domains, because they are already present. The increased intercalation capacity of the T1 phase, due to the smaller Ni : Mn ratio, can now be fully utilized from the start of the second cycle. On the basis of the fraction of the T1 phase at composition x ..., and the amount of lithium that can be inserted into it at the end of the first cycle ..., the capacity of the T1 phase in the second cycle evaluates to 0.74 mol of lithium, which is close to the observed 0.65 mol of the first plateau in the second cycle. ... This shows that in the second cycle most of the capacity of the T1 phase is utilized before the T2 phase is formed and that the first plateau has a larger capacity in the second cycle compared to the first cycle (0.35 mol of lithium)."

Perhaps the most important observation in this study is the disappearance of the initial Ni–Mn ordering due to the extensive migration of Ni and Mn, resulting in short-range order characterized by Ni-rich and Ni-poor domains. These clustering domains appear to exist over a length scale of approximately 40 nm, while the ordered domains have a length scale of approximately 13 nm. The resulting change in the electrochemical behavior signals the possibility of using transition metal clustering or migration as a method to engineer specific electrochemical performance. However, these particular spinel materials, which undergo the cubic to tetragonal phase transition, still exhibit significant volume changes during a phase change. It is noted in [16] that these significant volume changes are correlated with capacity decrease at increasing cycle number (Figures 5.11c and d) and the resulting mechanical strains within individual crystals and particles. The change from cubic symmetry to tetragonal symmetry with increasing Li content in Mn oxides can be attributed to the Jahn–Teller effect in the Mn cation environment, where the adjacent oxygen anions are distorted from an octahedral arrangement as the Mn valence approaches 3+. However, other transition elements such as Ni do not show the large distortions in the surrounding oxygen octahedra as the valence changes due to Li interaction. When Mn is the only cation species interacting with the oxygen sublattice and Li, this Jahn–Teller distortion can occur across many adjacent oxygen octahedra as the Li content increases. However, if the adjacent oxygen octahedra contain either Mn or Ni cations, then the Jahn–Teller distortion in the Mn octahedra may be inhibited by the stable octahedral symmetry of the Ni environment. This suggested interaction may provide a physical rationalization of the experimental observations. When $X = 1$, the Mn cations are ordered with the Ni cations. That is, there is a periodic arrangement (nonrandom) of Ni and Mn cations within the oxygen sublattice. This low-entropy arrangement signals a significant negative energy of interaction (attractive) between the Mn and Ni cations. However, as the Li content increases the Mn and Ni cations segregate to form clusters as the crystal structure changes symmetry from cubic to tetragonal (Figure 5.11e). This is another low-entropy arrangement, but now signals a large negative energy of interaction between Mn ions and a large negative energy of interaction between Ni ions. This dramatic shift in the energy of interaction is proposed to be a result of mechanical interaction between distorted oxygen octahedra. The idea embodied by this discussion can be quantified using regular solution theory applied to a ternary system. Let i,j,k correspond to Mn cation in an undistorted oxygen octahedra, Ni cation in an undistorted oxygen octahedra, and Mn in a distorted oxygen octahedral. Then the excess free energy of mixing is given as

$$G_{xs} = RT[(\chi_i RT \ln(\gamma_i)) + (\chi_j RT \ln(\gamma_j)) + (\chi_k RT \ln(\gamma_k))]$$

where γ_n is the activity coefficient and χ_n is the mole fraction for $n = i$, j, or k. According to the quasichemical model, the activity coefficient is correlated through the energy of interaction of the various chemical species as

$$RT \ln(\gamma_i) = (1 - \chi_i)(\Omega_{ij}\chi_j + \Omega_{ik}\chi_k) - \Omega_{jk}\chi_j\chi_k$$

where

$$\Omega_{jk} = ZN_o\left[H_{jk} - \frac{1}{2}(H_{jj} + H_{kk})\right]$$

The parameter Z is the number of neighboring octahedra, N_o is Avogadro's number, and H_{nn} is the bonding energy between neighboring octahedra and is expected to be a negative quantity in all cases. It is proposed that the ideas presented here in regards to clustering and ordering may be quantified using the quasichemical model when

$$|H_{ij}| > |H_{jj}| > |H_{jk}|$$

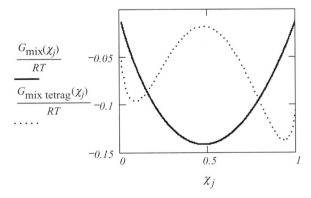

Ni octahedra mole fraction

Figure 5.12 Quasichemical solution models for the ordering (solid line) to clustering transition as the fraction of octahedra undergoing Jahn–Teller tetragonal distortion increases (dotted line).

and

$$H_{ii} = H_{jj} = H_{kk}$$

Now the energy of the system may be examined for the cases $X = 1$ and $X = 2$ in terms of the mole fraction of Ni octahedra. As shown in Figure 5.12, the energy of the crystal system, G_{mix}, is a minimum when the Ni octahedra have the same mole fraction as the undistorted Mn octahedra. However, when distorted octahedra are present ($X = 2$), the energy of the crystal system, $G_{mixtetrag}$, shows two local minimums. The double local minimum suggests that segregation of the Ni and Mn octahedra will be expected due to the difference in energy of interaction between the octahedra types.

5.4
Conclusion

The preceding discussion explores how it is possible to influence the voltage profile through variations in crystal chemistry. Once the atomic level interactions are understood, then the behavior during charge and discharge may be engineered to produce mechanically stable crystals with desired voltage and capacity properties. Basic concepts in materials thermodynamics coupled with mechanics of materials allow both a rationalization and development of design concepts for materials properties and structure relationships in battery electrodes. It should be acknowledged that many of the materials development in both oxides and phosphates were pioneered by J.B. Goodenough [17–19]. His insight and creativity have led to a new era in battery materials technology. The following chapters, which is concerned with anodes, will further illustrate how the capacity depends on the material selection and configuration.

References

1 Karthikeyan, K., Sikha, G., and White, R.E. (2008) *J. Power Sources*, **185**, 1398–1407.

2 Toomas Losin http://www.lenrek.net/experiments/nokia/ (accessed March 2008).

3 Yazami, R. and Ozawa, Y. (2006) *J. Power Sources*, **153**, 251–257.

4 Huang, S.H., Carter, C.W., and Chiang, Y.M. (2007) *Electrochem. Solid-State Lett.*, **10** (5), A134–A138.

5 Ramana, C.V. (2009) *J. Power Sources*, **187**, 555–564.

6 Chen, G., Song, X., and Richardson, T.J. (2006) *Electrochem. Solid-State Lett.*, **9**, A295.

7 Laffont, L., Delacourt, C., Gibot, P., Wu, M.Y., Kooyman, P., Masquelier, C., and Tarascon, J.M. (2006) *Chem. Mater.*, **18**, 5520.

8 Cahn, J.W. (1961) *Acta Metall.*, **9**, 795.

9 Cahn, J.W. (1962) *Acta Metall.*, **10**, 179.

10 Cahn, J.W. (1962) *Acta Metall.*, **10**, 907.

11 Gabrisch, H., Wilcox, J., and Doeff, M.M. (2008) *Electrochem. Solid-State Lett.*, **11**, A25–A29.

12 Thackeray, M.M., Johnson, C.S., Vaughey, J.T., Benedek, R., and Hackney, S.A. (2007) *J. Mater. Chem.*, **17**, 1–15.

13 Grey, C.P., Yoon, W.S., Reed, J., and Ceder, G. (2004) *Electrochem. Solid-State Lett.*, **7**, A29.

14 Tateishi, K., du Boulay, D., and Ishizawa, N. (2004) *Appl. Phys. Lett.*, **84**, 531.

15 Iguchi, E., Tokuda, Y., Nakatsugawa, H., and Munakata, F. (2002) *J. Appl. Phys.*, **91**, 2149.

16 Wagemaker, M., Ooms, F.G.B., Kelder, E.M., Schoonman, J., Kearley, G.J., and Mulder, F.M. (2004) *J. Am. Chem. Soc.*, **126**, 13526–13533.

17 Mizushima, K., Jones, P.C., Wiseman, P.J., and Goodenough, J.B. (1980) *Mater. Res. Bull.*, **15**, 783.

18 Padhi, A.K., Nanjundaswamy, K.S., and Goodenough, J.B. (1996) *Electrochim. Soc. Meet. Abstracts*, **96**(1), 73.

19 Padhi, A.K., Nanjundaswamy, K.S., and Goodenough, J.B. (1997) *J. Electrochem. Soc.*, **144**, 1188.

6
Next-Generation Anodes for Secondary Li-Ion Batteries
Katerina E. Aifantis

6.1
Introduction

This chapter will provide an overview of the materials chemistries and configurations that are being studied experimentally in order to develop the next-generation anodes for rechargeable Li batteries.

Until the 1980s, Li metal and alloys were used as the anode material [1], while various solid solution Li materials were used for the cathode. Due to safety problems that resulted from this materials selection, carbon-intercalation materials were solely substituted as the anode. The carbon structure, as well as the anode thickness, play a detrimental role in the number of Li ions it can intercalate [2]. Petroleum coke carbon, for example, which has a turbostratic structure, can intercalate a low amount of Li, forming Li_xC_6, where $0 < x < 0.5$ [3]. Graphite, however, which has almost a perfectly layered structure, is able to intercalate twice the amount Li, forming Li_xC_6, where $0 < x < 1$, and giving a capacity of 372 mAh/g [3]. During the first electrochemical cycles, the electrolyte decomposes on the carbon surface, forming a passivation layer (solid electrolyte interface, SEI); once this layer forms over the entire graphite surface, the electrolyte does not further decompose and the cell can be cycled continuously with no capacity loss [3]. Graphitic anodes are, therefore, used commercially in all Li-ion batteries. It should be noted that adding carbon black to graphite results in a better capacity retention [2], despite allowing for a lower Li intercalation. It has been possible to exceed the Li intercalation in graphite beyond LiC_6 [4a], but at the expense of safety issues, such as Li dendrite formation on the anode surface [4].

At the time that research was being performed to establish graphitic anodes, concurrent studies were being performed on metals that could host Li ions, without the occurrence of dendritic deposition of metallic lithium on their surface upon cycling [5]. These early papers suggested that Li intermetallics with various metals such as Sn, Si, Al, Bi, Pb, In, exhibited promising electrochemical properties [6–9]. The commercialization of these high-capacity anodic materials has not been possible, however, due to the large volume expansions, over 100% [5], that they experience upon lithiation. As a result during electrochemical cycling, the

High Energy Density Lithium Batteries. Edited by Katerina E. Aifantis, Stephen A. Hackney, and R. Vasant Kumar
© 2010 WILEY-VCH Verlag GmbH & Co. KGaA, Weinheim
ISBN: 978-3-527-32407-1

anodes expand (lithiation) and contract (delithiation) leading to fracture. This fracture produces active material that is no longer in electrical contact with the remainder of the electrode, and therefore becomes unable to respond to the applied voltages necessary to recharge or control the discharge of the battery. Furthermore, fracture increases the surface area available to chemical attack by corrosive agents present in the cell (HF and H_2O). Significant capacity drops are, therefore, observed after the first few cycles. Depending on the stability of the system, the surface cracks may lead to complete fracture of the electrode, thus resulting in safety issues as well.

In particular, the most promising materials for the next-generation anodes of Li batteries are Si and Sn, as they both allow for a maximum Li insertion that is over four times greater than that of graphite, $Li_{4.4}Si$ and $Li_{4.4}Sn$; while the respective specific capacities given off upon the formation of these Li-rich alloys are 4200 and 996 mAh/g, respectively [10], and their corresponding volume expansions are 300%. In this connection, it should be mentioned that the commercially used graphite has a very high mechanical stability as the volume expansions it undergoes are approximately 6–10%. In order to minimize the expansion of Si and Sn, nanoscaled Sn and Si anodes are being considered as nanomaterials not only have unique deformation mechanisms [11, 12], but also preferred electrochemical properties [13a].

It is well known that during charging of the battery a voltage is applied such that electrons flow from the cathode to the anode; as a result, the anode attains a negative charge and the positive Li ions from the cathode are attracted to it. Common electrolyte materials that accommodate the diffusion of Li ions, from one electrode to the other, are Li salts dissolved in an organic solvent (details on electrolyte solutions will be the focus of the following chapter). The exchange of Li ions during cycling affects anode materials significantly, both mechanically and chemically. This chapter will give an overview on the progress that has been made, thus far, in examining promising high-capacity, metal-based anodes. In addition to various materials chemistries, different configurations will be considered, as they also affect significantly the capacity of anodes.

6.2
Chemical Attack by the Electrolyte

Although studies existed since the 1970s [7, 8] suggesting that various metals such as Sn, Si, Bi, could intercalate Li, it was in the mid-1990s that industry began producing patents on promising such metal-based anodes that had the potential of replacing graphitic anodes; the proposed anodes would comprise of about 80% of the active metal-base powder, 10% carbon (for conductivity), and 10% PVDF (for connectivity). These patents by Fuji [14, 15] concerned anodes comprised of 5–10 μm powder particles of SnO. This material is also known as tin composite oxide (TCO). It provided a capacity of about 600 mAh/g [14].

The lithiation of TCO takes place in two steps. During the first electrochemical cycle, the Li ions react with the oxygen to form Li_2O, resulting in Sn nanoparticles dispersed in a Li_2O glass matrix [14]. Once this two-phase composite forms, the additional Li ions that diffuse into the anode, during the first cycle, react with Sn to form Li-rich Sn phases. During discharging, Li is extracted from the tin-rich phases, and then reinserted upon recharging. The lithium, however, that reacted with the oxygen to form Li_2O is not available to take place in further reactions in the cell [14] and this results in irreversible capacity loses observed in TCO materials, upon the first cycle. The aforementioned reactions can be written as [14–16]

$$SnO_2 + 4Li^+ + 4e^- \rightarrow 2Li_2O + Sn \text{ (irreversible reaction)}$$

$$xe^- + xLi^+ + Sn \cdot \leftrightarrow Li_xSn \ (0 \leq x \leq 4.4) \text{ (during continuous cycling)}$$

It was previously mentioned that electrochemical cycling of graphitic anodes resulted in the formation of a SEI layer on graphite; this protective layer inhibits co-intercalation of electrolyte solvents that result in exfoliation. In graphitic anodes, this SEI layer is formed once on the graphite surface during the first cycle (at voltages below 0.8 V Li/L^+) and its drawback is that it reduces the irreversible capacity. Such a layer also forms in these new TCO materials, but the sequence of reactions is not well clarified, and there are two possibilities:

1) The SEI layer forms first on the surface of the TCO, then Li ions diffuse through it, and decompose the TCO to Sn particles and Li_2O matrix; finally Li reacts with Sn to form rich Li–Sn alloys;

2) The Li reacts with the TCO to form the Li_2O and Sn particles, then the SEI layer forms on the surface of the particles, and finally the Li–Sn reaction takes place.

Impedance studies of TCO anodes [17] concluded that prior to the reaction of Li with O_2, Li first reacts with the surface of the material forming a SEI shell, which is distorted; then Li diffuses through this cell to distort the crystal structure of TCO and form the Li_2O matrix in which Sn is distributed, and then the reversible Li_xSn reaction takes place. The reaction during SEI formation is written as

$$Li^+ + (EC, DEC, ...) + e^- \rightarrow SEI \ (Li_2CO_3, ROCO_2Li, ...),$$

where (EC,DEC, ...) denote the electrolyte components.

The composition of the SEI layer has been identified for TCO anodes as Li_2CO_3, which forms at 1.2–0.9 V Li/Li^+, and $ROCOLi_2$, which forms later at voltages below 0.9 V Li/Li^+; it follows, therefore, that the Li_2Co_3 forms first [17]. The SEI layer in TCO is a greater problem than it is in graphitic anodes, since as the Sn sites fracture during cycling, due to the expansions they experience, more surface is available to chemical attack by the electrolyte. Furthermore, as the sites expand and contract, upon Li insertion and deinsertion, the SEI layer forms and disintegrates continuously, and therefore new SEI layers form on aged layers, and eventually a thick layer forms that inhibits Li diffusion in the active material, and reduces the battery capacity. A problem in studying this electrochemical damage through microscopy is that it is difficult to distinguish it from the mechanical damage induced by the volume expansions.

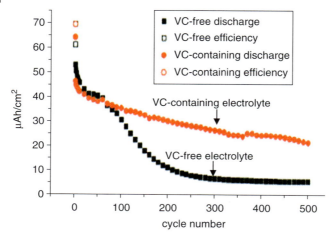

Figure 6.1 Si thin films cycled in a VC-free and VC-containing electrolyte. (Reproduced from Ref. [20].)

Li intercalation and SEI layer formation are similar in other metal-based anodes, and studies are being performed to fully understand this process. In trying to minimize the SEI thickness, studies are being performed to estimate the operation voltages that inhibit such electrolyte decomposition; in [18], for example, it is indicated that the formation of this layer on the surface of AlSb anodes is reduced at voltages below 0.5 V.

Another way to face this problem is by considering alternate electrolytes. It was first shown that adding vinylene carbonate (VC) in electrolytes enhances the performance of graphitic anodes, by reducing its irreversible capacity and improving its cyclability [19]. This led to the recent study [20] in which the same thin film Si anodes were cycled using an electrolyte that contained VC and one that did not. In the VC-free electrolyte case, the SEI thickness kept increasing with cycling, and it contained a higher content of compounds, such as LiF, that decrease the capacity. As a result, the VC-free electrode had an increased anode polarization, which resulted in the degradation of cycle performance. The VC-containing electrolyte, however, allowed for an SEI layer to form that had a stable thickness and, therefore, a higher capacity was retained for 500 cycles for the VC-containing electrolyte [20], as shown in Figure 6.1. New electrolytes that are being investigated for Li batteries will be described in detail in the next chapter.

6.3
Mechanical Instabilities during Electrochemical Cycling

Although the formation of the SEI layer is of importance, it does not play such a detrimental role in capacity fade as fracture. In Figure 6.1, it was illustrated that controlling the formation of the SEI layer, through the electrolyte, resulted in a higher capacity. It can be seen, however, in Figure 6.1 that for both types of electrolytes a significant capacity drop took place, regardless of the structure of the passivation layer. This capacity drop resulted from the 300% expansion of the Si thin film, which leads to fracture. In continuing the study of these next-generation anodes, it is essential to describe the mechanical instability phenomena that arise upon electrochemical cycling (a brief description of how candidate electrode materials are cycled was given in Chapter 3).

As the Li ions enter the anode, they diffuse through the inactive regions that may be present and they enter the active with respect to Li metal inclusions (Si, Sn, Al, Sb, Bi can act as active sites). As Li enters the active sites, the metal expands over 200%. During discharge of the battery, that is, during operation, electrons flow back into the cathode and hence the Li ions are attracted back to their original positions. As a result, the active sites in the anode are free to attain the initial volume they had – prior Li insertion – and a contraction takes place, leading to fracture.

This cracking forms during the initial electrochemical cycles, and has been documented very nicely in [10]. In [10], a 4-μm Cu film, which is inactive with respect to Li, was sputtered on a stainless steel substrate, and then a 1-μm SiSn thin film was sputtered on Cu. As a complete battery system was not examined in [10], but instead the behavior of the SiSn film upon Li insertion and deinsertion was considered, the authors of [10] refer to the addition of Li in SiSn as the discharge process, while the removal of Li is referred to as the charging process; pure Li metal was used as the counter electrode. Upon Li insertion, Si and Sn experience an expansion of 310% and 260%, respectively. To determine the direction of this expansion, razor marks were carved on the SiSn film prior the initiation of electrochemical cycling (Figure 6.2a). Figure 6.2b was taken after the first

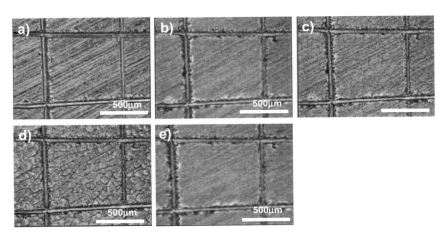

Figure 6.2 Experimental evidence taken from Ref. [10] on electrochemical cycling of a Si–Sn thin film. (a) Before cycling begins, (b) after first Li insertion, (c) 1 hour during first Li deinsertion, (d) after complete Li deinsertion, and (e) after second Li insertion. (Reproduced from Ref. [10].)

discharge (i.e., after SiSn was fully lithiated); it can be seen that no lateral expansion is observed and, therefore, the film expanded in the out-of-plane direction. In Figure 6.1c, it can be seen that as Li is removed, from SiSn, cracks begin to form, which become more apparent as more Li is removed during continuous charging, as seen in Figure 6.2d. The flake-like active particles produced from these cracks are 100 μm. These particles have the ability to continuously expand and contract upon cycling and it can be seen that once the film becomes fully lithiated (Figure 6.2e), the cracks are not visible and the film looks as it did during the first discharge (Figure 6.2b), before cracking took place. It was observed at higher resolution images that in fact the cracks do remain, but the defined form of these cracks particles can expand and contract without further cracking. Therefore, even though SiSn is fractured it is possible to maintain electrical integrity and control the charging and discharging processes; this electrical contact results from the fact that the center of the cracked particles remains pined to the Cu film and, therefore, electrical contact is sustained.

This submicroscopic fracture of the SiSn anode that results from electrochemical cycling resembles the macroscopic phenomenon of drying mud, as seen in Figure 6.3; it is very interesting that the pattern in both the cracked active site and cracked mud are similar, regardless of the great scale difference. In this case, Li acts in a similar way as the evaporating water; in fact just as dry mud, the edges of the cracked SiSn particles curl at the edges, where they detach from the Cu film.

It has been possible for the 30-μm diameter flake-like active particles with 1–8 μm thickness to expand and contract by up to 100% with no noticeable damage, after the first discharge process [10]. Similar cracking behavior has been observed when electrochemically cycling an SiO$_2$ thin film [21]; the cracked particles produced, however, had a much smaller diameter of 100 nm. It should be noted here that it has been proposed that these cracks might act as high diffusivity paths for the Li ions, allowing for better electrochemical properties of the anode through some cracking; this has not been examined in depth though.

(a)   (b)

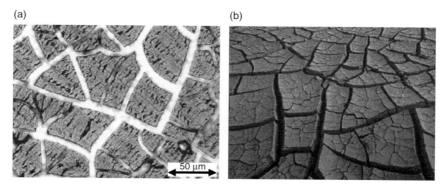

Figure 6.3 (a) An optical micrograph of a Li alloy film after expansion and contraction due to electrochemical cycling. (b) Cracked mud in a dry lake bed bottom. (Reproduced from Ref. [10].)

In the remainder of this chapter, various methods that have been developed to deal with the aforementioned instability phenomena of these promising anodes will be described.

6.4
Nanostructured Anodes

Extensive experimental research has yielded that the most effective way to minimize the large expansions of metals that are active with respect to Li is to surround them by an inactive or less active material with respect to Li. As a result, the expansion upon Li insertion is both constrained and buffered. Furthermore, these active/less-active composite materials are taken to have nanodimensions since deformation and fracture mechanisms are less severe at the nanoscale [11, 12], and therefore the anode capacity does not decrease as rapidly with continuous electrochemical cycling. In particular, crumbling of the active site surface during cycling is reduced at the nanoscale, and the concentration gradients responsible for fracture are significantly minimized. Pulverization of the active materials is also reduced at the nanoscale; however after continuous cycling nanoparticles aggregate, under the absence of a less-active matrix, and therefore pulverize.

In addition to the mechanical stability, the electrochemical properties are also enhanced at the nanoscale (i) due to the large surface area present in nanomaterials, higher charge/discharge rates are achieved, (ii) nanodimensions result in shorter path lengths for Li-ion transport, that is, higher diffusivities, allowing for an increase in power capabilities; for example, bulk Si reacts with Li at 400 °C [13b], whereas nanoscale Si reacts with Li at room temperature [13c].

The most effective way to deal with both the electrochemical (SEI layers) and mechanical (fracture) instabilities that arise at the nanoscale is to embed the active material, which can be Si, Sn, Sb, Si, Al, in an inactive or less active with respect to Li matrix, such as Cu, N, Ag, C; or furthermore to encase the active sites in a carbon shell. As previously mentioned, this configuration minimizes damage and fracture of the active sites since the matrix material buffers (cushions) and constrains the expansion of the active material. Furthermore, the active site surface area (which is significantly high at the nanoscale) is protected by the matrix, and therefore a thick SEI layer does not form on the active site surface. Finally, the matrix inhibits agglomeration and pulverization of the active materials. In the cases where C is the matrix, in addition to offering mechanical and electrochemical stability, it allows for additional storage of Li ions.

There exist numerous experimental studies that propose promising materials selections that allow for capacities nearly double that of current graphitic anodes. In particular, these configurations exist in various configurations such as (i) thin films, (ii) fiber-like/nanowires, (iii) active/less-active nanocomposites. In the sequel, the various next-generation anodes will be presented in the aforementioned three categories. Each configuration category will give information on the fabrication, microstructure, and electrochemical performance of various promising Sn-based and Si-based anodes as they offer the highest capacities and are, therefore, the most examined. In the last section, Sb, Al, and Bi anodic materials will also be summarized, although they appear to be the furthest away from commercialization.

6.5
Thin Film Anodes

6.5.1
Sn-Based Thin Film Anodes

Although Sn is not the material that gives the highest capacity upon its reaction with Li, it will be examined first since it was the metal that popularized the use of these high-capacity metal-base anodes, after the patent by Fuji Film [15]. It should be noted that pure Sn is hardly used alone as an anode, but almost always in its composite form, mainly as SnO_x, since the Li_2O matrix that forms upon Li insertion buffers/constrains the expansion of Sn, which is approximately 298%.

In [22], 550 nm films were fabricated by electroplating SnO_2 on a Pt current collector. The resulting electrodes had an initial capacity of 560 mAh/g, which reduced to 440 mAh/g after 40 cycles.

More efficient Sn-based thin film anodes can be achieved by using a different matrix material. It was observed in [23] that Li could form a buffering matrix with S, similarly as it does with O. Studies performed in [24] propose that the Li intercalation takes place as

$$x\text{Li}^+ + \text{SnS} + x\text{e}^- \rightarrow \text{Li}_x\text{S} + \text{Sn} \text{ (irreversible reaction)}$$

$$x\text{e}^- + x\text{Li}^+ + \text{Sn} \leftrightarrow \text{Li}_x\text{Sn} \, (0 \leq x \leq 4.4) \text{(during continuous cycling)}$$

In [25], SnS/C thin films anodes were fabricated by homogeneously dispersing 5–6 nm SnS particles into carbon aerogel and then coating this thin film on copper foil through the spin-coating method. Carbonizing the mixture resulted in an SnS/C nanocomposite thin film that was 600 nm. Therefore, C provided additional constraints to the expansion of Sn, and also since C is active with respect to Li, it could store additional Li^+. The initial capacity of this thin film was nearly 1000 mAh/g, while after a few cycles it stabilized at 530 mAh/g for 40 cycles. The significant capacity drop of about 500 mAh/g that occurred during the first cycle (Figure 6.4) was due to the irreversible bond of Li ions to form the Li_xS matrix. Once this matrix formed the capacity was stable, and the presence of C and S prevented aggregation and fracture of the Sn particles (Figure 6.5), and minimized the formation of the SEI layer; the small dimensions of the Sn particles also play a detrimental role in the minimal damage observed.

There exist additional Sn-based thin films, to those described above; however they give capacities (below 400 mAg/h), which are comparable to those of commercial carbonaceous anodes and therefore will not be discussed.

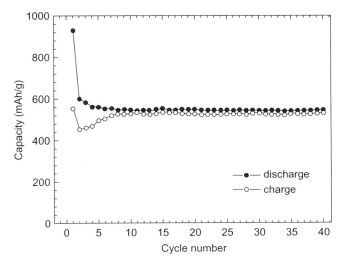

Figure 6.4 Capacity retention of SnS/C nanocomposite thin film. (Reproduced from Ref. [25].)

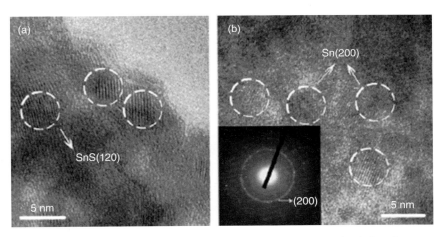

Figure 6.5 SnS/C nanocomposite film. The left HRTEM micrograph is taken before the electrochemical cycling begins, while the left is taken after the fourth electrochemical cycle. It can be seen that no significant damage has taken place. (Reproduced from Ref. [25].)

6.5.2
Si-Based Thin Film Anodes

The other most popular anode candidate for Li batteries is Si. As opposed to Sn, which is rarely used in its pure form, pure metal Si has been used as an experimental anode base material, especially in the film configuration. Theory predicts that the formation of $Li_{4.4}Si$ produces a capacity of 4200 mAh/g. The capacity of bulk Si is reduced by 90% after the fifth electrochemical cycle [26], such that it approaches the capacity of C. For nanoscaled Si, however, it has been possible to maintain capacities close to the theoretical value. In particular, the highest capacities have been obtained for Si in the form of films with nanodimensions. The focus of researchers has been to increase the film thickness, while maintaining high capacities over 1500 mAh/g. It has been observed that the capacity not only depends on the film thickness, but also on the fabrication method. Therefore, an overview will be given here on the various capacity retentions that have been achieved in various studies. It should be noted that the common factor in all the proceeding studies is that Si films have an amorphous structure instead of a crystalline, as it has been documented that amorphous Si has a more open structure as compared to crystalline Si, and can therefore accommodate more efficiently the Li insertion and deinsertion by allowing for a homogenized expansion–contraction upon cycling [27].

In [27], it was shown that a capacity of 2000 mAh/g for over 50 cycles can be achieved for an amorphous 100 nm Si thin film fabricated by physical vapor deposition. In [28], it was observed that both the film thickness and capacity can be increased by fabricating the films through radio frequency magnetic sputtering. This allowed the development of 250 nm amorphous Si thin films that gave capacities of 3500 mAh/g for 30 cycles [28]; increasing the film thickness to 500 nm, while maintaining this fabrication method (but using a rough Cu collector), gave capacities of 1500 mAh/g after 30 cycles [29]. In [30], a capacity of 3000 mAh/g for 70 cycles was achieved for 200 nm Si thin films. This radio frequency magnetic sputtering technique, however, significantly reduces the favorable properties of Si, when used for the production of microscale films [28], and as seen by the aforementioned it has not been possible to overcome this. Therefore, despite the significantly high capacities obtained, it is not likely that this method will allow for the development of commercial Si anodes.

In [31], it was possible to increase the film thickness, while maintaining a high capacity by roughening the surface of the substrate (Ni) onto which n-type Si was deposited in a vacuum. Roughening of the Ni surface substrate was performed by etching. Letting the thickness of the Si film be 500 nm allowed capacities of 3600 mAh/g to be retained for 200 cycles [31].

In attempts to find alternative fabrication methods that allow the production of thicker films that retain the favorable properties of Si, electron-beam deposition was used to develop 2-μm Si films [32]. These films were able to deliver a 3400 mAh/g capacity for 50 cycles, which decreased to 2550 mAh/g after 200 cycles [32]. Motivated by these results, a battery cell was developed [32] in which the anode was a Si thin film with 4–6 μm thickness, while the cathode consisted of $LiCoO_2$; this cell retained a capacity of 1.5 mAh/cm^2 for 200 cycles, with an efficiency of 97%. It is interesting to note the microstructure effects that Li insertion had in the Si film anode. Figure 6.6 illustrates the as-deposited Si film on the roughened Cu substrate. Due to the roughness of Cu, the Si film has a columnar structure. These columns in the Si film result in a discontinuous structure, which as a result contains additional Si interfacial surface, perpendicular to Cu. The discontinuous

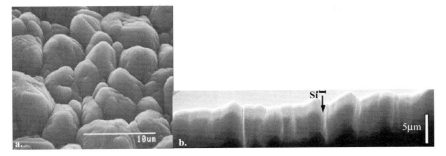

Figure 6.6 A 6-μm Si film anode, prior cycling: (a) SEM image of the anode surface, (b) TEM image of the cross-section. (Reproduced from Ref. [32].)

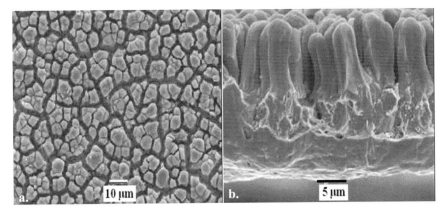

Figure 6.7 SEM images of a 6-μm Si film anode after 250 cycles (produced capacities of 1.8 mAh/cm^2): (a) surface, (b) cross-section. It can be seen that Li insertion and deinsertion in the first cycles resulted in Si-micron fibers/columns. (Reproduced from Ref. [32].)

structure accommodates the Si expansion upon Li insertion, while the perpendicular interfaces allow for the fracture to form along the interface, and therefore the microcolumn structure is accentuated upon cracking during discharge. Planar fracture is, therefore, avoided. Figure 6.7 depicts the aforementioned microcolumn structure after 250 cycles. This fracture morphology most likely occurs during the initial electrochemical cycling. Just as the flake-like particles of Figure 6.2, the Si columns remain in contact with Cu. It can be seen that ample space exists to accommodate the volume expansions, and the Si microcolumns remain in contact with the Cu substrate. This good adhesion is achieved due to the roughness of Cu [32].

In addition to the aforementioned LiCoO$_2$/Si thin film cells, another type of Si thin films were cycled with an LiCoO$_2$ cathode [33]. The Si film was sputtered on rough Cu foils by dc magnetron sputtering; the resulting films were 2 μm thick, and when cycled against an LiCoO$_2$ cathode, a capacity of 0.54 mAh/cm^2 for 300 cycles was achieved, with only a 0.01% capacity decay per cycle. Figure 6.8 depicts

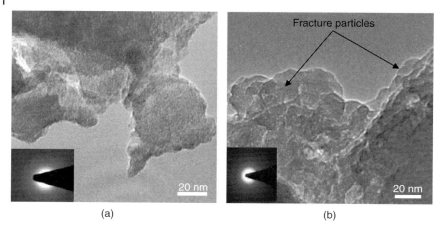

Figure 6.8 (a) HRTEM micrograph of an amorphous 2-μm Si thin film deposited on Cu. (b) HRTEM micrograph of the Si thin film after 300 cycles. (Reproduced from Ref. [33].)

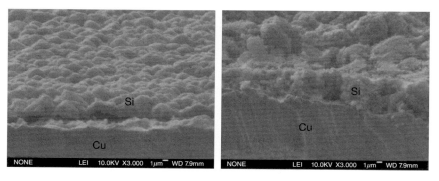

Figure 6.9 (a) SEM micrograph of Si thin film (2 μm thick) deposited on Cu before cycling. (b) SEM micrograph of Si thin film after 300 cycles (the thickness has increased to 6 μm). (Reproduced from Ref. [33].)

the amorphous structure of Si (which allows for a better accommodation of the volume expansions) prior cycling, and the pulverized Si film, with increased thickness, after 300 cycles. Apparently, however, this fracture is not severe, as the capacity retention was very good, implying that the fracture particles remained in contact with the Cu, and could respond to electrochemical cycling. It is of interest to point out that the formation of the SEI layer can be significant on such Si films, as shown in Figure 6.9. The capacity was not reported to be affected by the SEI layer [33]; in [32], it is reported, however, that Li ions were captured in this layer and resulted in an increase in the irreversible capacity.

Before concluding the thin film configuration, it is interesting to note a recent study that emphasizes the preferable effects that the addition of N has in Si thin films [34a]. Amorphous $Si_{1-x}N_x$ thin films were deposited on a Cu substrate through

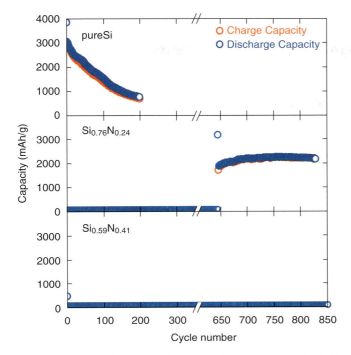

Figure 6.10 Electrochemical performances of various 200 nm $Si_{1-x}N_x$ thin film anodes. (Reproduced from Ref. [34].)

rf magnetron sputtering. The resulting films had a 200 nm of thickness. Initially, as Li enters the film three phases form Li_3N, Si_3N_4, and Si, where the Si active sites are distributed throughout. In addition to offering mechanical stability, the matrix formation enhances the electrochemical properties, as Li_3N is one of the best Li-ion conducting compounds. In order to study the effect of N, three different thin films were examined: a pure Si thin film, an $Si_{0.73}N_{0.24}$ film, and an $Si_{0.59}N_{0.41}$ film [34a]. Each anode exhibited a very different electrochemical performance, which is seen in Figure 6.10. The pure Si anode in this case exhibited a high initial capacity of 3000 mAh/g, which decreased to 700 mAh/g at the 200th cycle. The $Si_{0.73}N_{0.24}$ anode exhibited practically no capacity for the first several hundred cycles, while an abrupt capacity increase of nearly 2000 mAh/g was recorded at the 646th cycle, which increased to 2300 mAh/g at the 750th cycle, and was retained until cycling ceased at 850 cycles. The $Si_{0.59}N_{0.41}$ anode did not give off any capacity throughout the whole cycling process. The unique electrochemical behavior of $Si_{0.73}N_{0.24}$ is attributed to the formation of Li_3N, which began forming at the 646th cycle and, therefore, Si particles were created that could react with Li; further cycling formed additional Li_3N, which enhances the kinetics of Li insertion and deinsertion, and therefore the capacity increased as shown in Figure 6.10. Furthermore, it was observed in [34a] that performing cycling at higher temperatures allowed the abrupt capacity increase for an $Si_{0.73}N_{0.24}$ anode to occur around the 30th cycle, since high temperatures aid the formation of Li_3N.

As an end note, it should be illustrated that in addition to adding N, other materials can be used as matrices in Si-based thin films. Si nanoparticles (30–50 nm) were distributed homogenously in $LiTi_2O_4$ [34b] using a sol–gel method. The resulting nanocomposite film was porous, accommodating therefore expansions, and had a thickness of 600 nm. A capacity of 1100 mAh/g was retained for 50 cycles. Through appropriate impedance measurements, it was possible to measure the Li-ion diffusion coefficient as 10^{-13} cm^2/s.

The deficiency of using thin film electrodes, though, is that their thickness does not provide sufficient active material to make them commercially viable. Hence, alternative nanoconfigurations are sought after.

6.6
Nanofiber/Nanotube/Nanowire Anodes

It was seen in the previous section that the most promising thin film configuration was that in which the Si thin film had a columnar structure, since it could accommodate fracture through the formation of microncolumns (Figures 6.6 and 6.7). In this section, it will be further illustrated that such type of vertical/columnar-like microstructure results in higher capacities not only for Si- but also for Sn-based anodes.

6.6.1
Sn-Based Nanofiber/Nanowire Anodes

The highest stable capacities that have been achieved, thus far, for Sn-based anodes are those obtained for the fiber-like Sn-based materials presented in [22]. In Section 6.5.1, the results for the SnO_2 thin film described in [22] were reported, and as seen they did not have a high capacity retention. In addition, however, to a thin film SnO_2 anode, the authors in [22] examined a fiber-like SnO_2 anode. These fiber-like materials were produced similarly as the thin film, but on performing the template method a microporous membrane was placed on the Pt current collector, and therefore the SnO_2 precipitated in the pores and formed fibers that protruded from the current collector as seen in Figure 6.11. The nanofibers were 110 nm in diameter. Their initial capacity was 700 mAh/g, which increased after the first few electrochemical cycles and stabilized at 770 mAh/g. It is possible to cycle such anodes 800 times without a decrease in capacity. The unique electrochemical performance of these anodes is attributed to their nanodimensions and also to the fact that the brush-like fibers accommodate the expansion during Li insertion. For illustration purposes, the capacity of the SnO_2 thin film is also shown in Figure 6.12.

Another confirmation of the preferred electrochemical properties of nanofibers, was presented in [35]. SnO_2 nanorods that can be used as anodes were fabricated

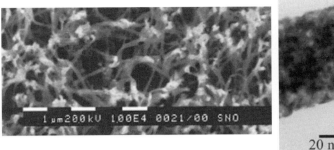

Figure 6.11 Fiber-like SnO_2 nanostructured anodes. (Reproduced from Ref. [22].)

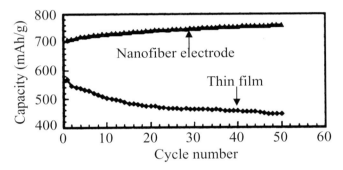

Figure 6.12 Comparison contained in [22] of capacity retention between SnO_2 nanofiber and SnO_2 thin film electrodes, at a charge/discharge rate of 8 C over the potential window of 0.2–0.9 V. (Reproduced from Ref. [22].)

by heat-treating Sn–20%Ag alloys in a gas containing oxygen. If the oxygen volume is between 0.5 and 1%, SnO_2, nanowires (30–70 nm in diameter and several hundred micrometers in length) form on the Sn–Ag surface as shown in Figure 6.13. If the oxygen volume level, however, is above 2%, nanoparticles form that have a mean diameter of 30 nm; nanowires are still sparingly observed in this case. As Figure 6.13 indicates, the SnO_2 powder particles agglomerate, not leaving much room to accommodate the volume expansion upon cycling; the SnO_2 nanorods, however, allow for better mechanical stability and therefore a significantly higher capacity retention as seen in Figure 6.14.

6.6.2
Si-Nanowire Anodes

As mentioned in Section 6.4, graphite is often used to coat active materials so as to protect the surface and accommodate the volume expansion. Therefore, as Li_2O acted as the buffering matrix in the aforementioned Sn-based anodes, the authors

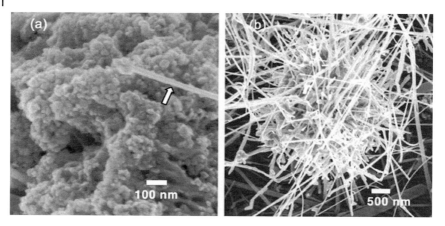

Figure 6.13 (a) Nanopowder of SnO$_2$ agglomerate particles formed by heat treating Sn–Ag alloy in the presence of a greater than 2% oxygen volume; a SnO$_2$ nanowire is indicated (b) SnO$_2$ nanorods/nanowires formed by heat treating Sn–Ag alloy in the presence of a 0.5–1% oxygen volume. (Reproduced from Ref. [35].)

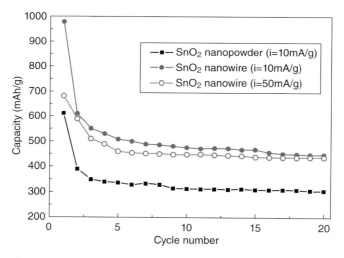

Figure 6.14 Superior capacity retention of SnO$_2$ nanorods over SnO$_2$ nanoparticles. (Reproduced from Ref. [35].)

in [36] used C to protect the surface of Si nanowires. Since carbon nanotubes have been given considerable attention during the past decade, Cr-doped Si nanowires were coated by C nanotubes as seen in Figure 6.15. Cromium was added in order to increase the conductivity of Si (and hence assist Li diffusion). The nanowires were coated by C nanotubes through decomposition of CH$_4$ by using Ni as the catalyst at a temperature of 1073 K for 6 h. The initial capacity was 1700 mAh/g, which reduced to 1250 mAh/g after the tenth cycle that was stably sustained.

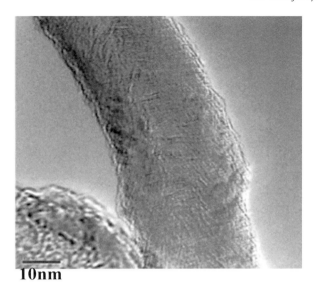

Figure 6.15 Cr-doped Si, coated with carbon nanotube. (Reproduced from Ref. [36].)

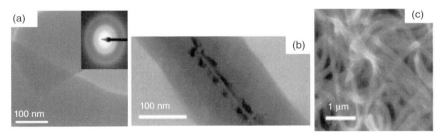

Figure 6.16 (a) TEM image of Si core/shell NWs after third electrochemical cycle, (b) TEM image after 15th cycle, (c) SEM image after 15th cycle. (Reproduced from Ref. [37].)

Promising capacities have also been obtained for Si nanowires. Although these nanowires are solely comprised of Si, they form core/shell structures as Si inside the nanowire is crystalline, while it is surrounded by amorphous Si [37]. The amorphous Si has a lower activation voltage than the crystalline Si; hence by electrochemical cycling at particular voltages it is possible to activate only the amorphous Si shell so as to accept Li ions. The crystalline Si core allows for high electrical conductivity and mechanical support of the nanowire, while the amorphous shell can store Li ions and accommodate the resulting volume expansions. The capacity attained for these anodes was as high as 1000 mAh/g for 100 cycles, while the coulombic efficiency was 90% [37]. Figure 6.16 depicts the crystalline/ amorphous core/shell Si nanowire, after electrochemical cycling. The dark and light contrast regions of the core, after the 15th cycle, indicate the straining that took place upon cycling due to the volume expansions.

6.7
Active/Less Active Nanostrutured Anodes

As was seen from the previous section, Sn- and Si-based anodes allow for higher capacities to be obtained than from those produced by carbonaceous materials. Although thin film and nanowire configurations allow for very good electrochemical properties, more attention has recently been given to nanostructured composites, in which a more active material is alloyed or embedded in a less-active material with respect to Li.

6.7.1
Sn-Based Active/Less Active Anodes

6.7.1.1 Sn–Sb Alloys

In addition to binding Sn to O, Sn–Sb have also been proposed as anodes. Although Sb is used to buffer the volume expansion of Sn, in a similar way as the use of O, it is also active with respect to Li, providing a capacity of 660 mAh/g, upon maximum lithiation (Li_3Sb).

To show the enhanced electrochemical properties that result from alloying Sn with Sb, various Sn–Sb alloys were considered as anodes in [38]. It was noted that the capacity of pure Sn nanoparticles (diameters between 50 and 200 nm, Figure 6.17a) is initially 990 mAh/g but drops rapidly to 210 mAh/g after 20 cycles; adding, however, 46.5at%Sb (Sn–Sb particle diameters between 50 and 200 nm, Figure 6.17b) results in anodes that give an initial capacity of 701 mAh/g, which reduces to 650 mAh/g after 20 cycles. Other at%Sb compositions (Sn-30.7%Sb, Sn-46.5%Sb, Sn-47.2%Sb, Sn-58.5%Sb, Sn-80.8%Sb) also result in a better cyclability than pure Sn, but 46.5at%Sb gave the best capacity retention [38]. It follows that since Sb provides a lower capacity than Sn, increasing the amount of Sb in the Sn–Sb alloy lowers the initial capacity; this side effect, however, is greatly compensated by the significantly increased capacity retention of Sn–Sb alloys.

In order to further enhance the enhanced cyclability of Sn-based anodes, there have been studies that coated the Sn–Sb alloy with a C shell. In [39], the cyclability of a crystalline Sn_2Sb micron-powder, Figure 6.17c, was compared to the same powder but with the Sn_2Sb particles encapsulated in C microspheres, Figure 6.17d, through carbonization. The starting reversible capacity of the 5–10 μm Sn_2Sb particles was 689 mAh/g, but a significant capacity fade was observed after the 13th cycle (approximately), and after the 60th cycle the capacity had reduced to 150 mAh/g; even lower than that of graphite. Once, however, these powders were encapsulated in C, the capacity was retained to 600 mAh/g after 60 cycles, Figure 6.18.

It is noted here that the Sn-46.5at%Sb anodes [38] do not show a significant capacity loss at 13 cycles as the Sn_2Sb anodes [39]. This is attributed to the fact that the Sn-46.5at%Sn particles are at the nanoscale, whereas the Sn_2Sb have a micron size. As a result, the deformation mechanisms of the Sn-46.5at%Sb anode are less severe, and therefore agglomeration and subsequent pulverization are observed at much slower rates than in the micron-scale. A most effective SnSb anode would, therefore, result by encapsulating nanosized Sn_2Sb powder by C microspheres.

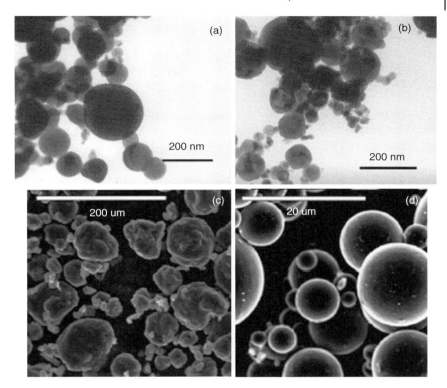

Figure 6.17 (a) Sn nanoparticles (reproduced from Ref. [38]); (b) Sn-46.5at%Sb (reproduced from Ref. [38]); (c) Sn₂Sb micron-powder particles of multiangular shape (reproduced from Ref. [39]); (d) Sn₂Sb particles encapsulated in C microspheres (reproduced from Ref. [39]).

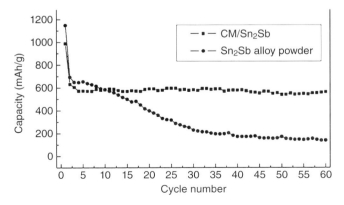

Figure 6.18 Comparison of capacity retention between Sn₂Sb alloy power and carbon-coated Sn₂Sb powder. (Reproduced from Ref. [39].)

6.7.1.2 **SnS₂ Nanoplates**

Another promising Sn-based active/less active anode is that of layered SnS_2 nanoplates. The fabrication of these unique nanostructures is achieved by thermal decomposition of the precursor, $Sn(S_2CNEt_2)_4$, in an organic solvent at elevated temperature [40]. These nanoplates are 120 nm in the lateral direction, and they are stacked on each other along the [001] direction; their stacking configuration is enforced by van der Waals interactions [41]. Each nanoplate is single crystalline and is comprised by 30 layers according to Figure 6.19, while their shape is hexagonal as depicted in Figure 6.20.

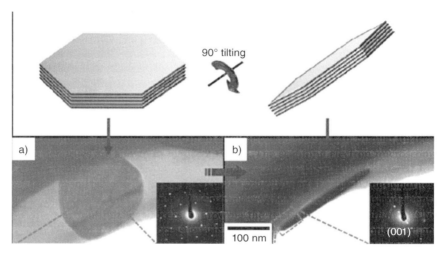

Figure 6.19 TEM images, obtained by rotating the TEM holder stage 90°, indicating the layered structure of the SnS₂ nanoplates. (Reproduced from Ref. [41].)

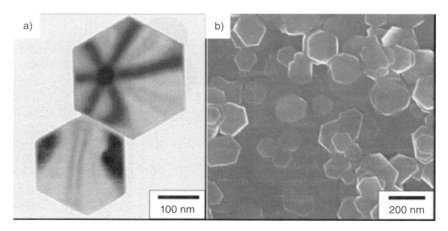

Figure 6.20 (a) TEM and (b) FESEM images of SnS₂ nanoplates. (Reproduced from Ref. [41].)

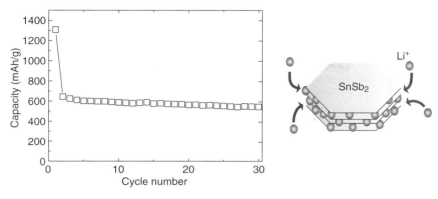

Figure 6.21   Capacity retention of SnS$_2$ nanoplates, Li insertion in SnSb$_2$ nanoplates. (Reproduced from Ref. [41].)

Upon Li insertion during electrochemical cycling, the lithiation that takes place is as follows:

$$SnS_2 + 4Li^+ \rightarrow 2Li_2S + Sn \text{ (irreversible reaction)}$$

$$xLi^+ + Sn \leftrightarrow Li_xSn \ (0 \leq x \leq 4.4) \text{ (during continuous cycling)}.$$

The initial capacity of these materials was 645 mAh/g, which stabilized to 583 mAh/g, as seen in Figure 6.21. The high capacity retention of these SnS$_2$ nanoplates is attributed to the following factors: (i) the Li$_2$S matrix buffers the expansion of Sn, and (ii) their 2D layered structure results in a finite lateral size with enhanced open-edges. As a result the Li-ion diffusion through the active materials is increased, while the overvoltage associated with the formation of Li–Sn is decreased [42].

6.7.1.3 Sn–C Nanocomposites

In concluding, the Sn-based active/less-active nanostructured anodes, Sn–C nano-composites will be described. In [43], a fabrication method by which Sn nanoparticles were attached on the surface of a synthetic carbon was proposed; however, the resulting capacities were not above 400 mAh/g. In [44], a similar fabrication method was presented but the resulting Sn/C composite contained Cl, which is corrosive and its presence should be avoided in battery cells. In [45], a stable capacity of 500 mAh/g was attained for 25 cycles; however, in addition to Sn particles that were attached on the C surface, free-standing Sn particles were also present; these free-standing particles agglomerated and fractured, and hence were not able to contribute in the capacity of the anode. In [46], however, a fabrication method was developed for binding all the Sn particles on the C surface; in some cases, Sn was oxidized and hence SnO$_2$ was bound on the C surface. As seen in Figure 6.22, the smallest Sn or SnO$_2$ particles were 4 nm, while in some cases they aligned

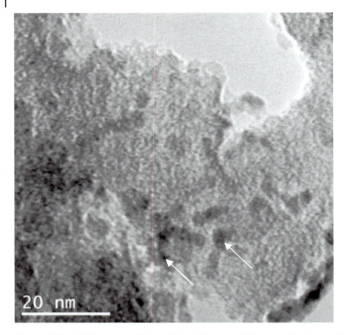

Figure 6.22 TEM image of Sn or SnO$_2$ islands on Vulcan C. The Sn or SnO$_2$ islands are shown as dark spots such as those pointed to by the arrows. The Sn content is 8 wt.%. (Reproduced from Ref. [46].)

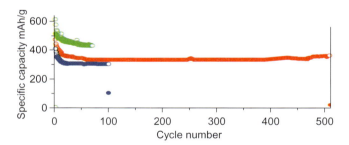

Figure 6.23 Capacity retention at various current densities for 8 wt.% SnC nanopowder (again Vulcan C was used). Green: 0.2 A/(cm^2 g$_{AM}$); Red: 1.025 A/(cm^2 g$_{AM}$); Blue: 1.011 A/ (cm^2 g$_{AM}$). (Reproduced from Ref. [47].)

adjacent to each other forming larger diameter Sn or SnO$_2$ islands; hence this microstructure is also referred to as Sn/SnO$_2$ island microstructure. In Figure 6.23, it can be seen that a stable capacity of 380 mAh/g was attained for over 400 cycles for such a 8 wt.%Sn–C island composite. Although this capacity is slightly higher than that of currently used graphitic anodes, it should be noted that the carbon matrix was Vulcan-C, which provides a capacity or 180 mA/g; therefore, the 200 mAh/g capacity increase observed was solely due to the addition of 8 wt.%Sn. Current studies are being performed with high capacity graphites.

It should be mentioned that Sn–C nanocomposites have also been fabricated in which the Sn nanoparticles are embedded in C (e.g., [48, 49]). When the Sn particles were over 100 nm [48], the capacities were not as high; if the Sn nanoparticles, however, had an average diameter of 35 nm, a capacity of 500 mAh/g was retained for 100 cycles. To achieve this capacity, the Sn content had to be 50 wt.%, whereas the aforementioned Sn/C island composites allow for higher capacities with a lower Sn content [45].

6.7.2
Si-Based Active/Less Active Nanocomposites

Initial studies for Si-based anodes in a nanoparticle configuration employed pure Si, solely, just as was done for the thin-film configuration. The achieved capacities, however, were not only lower than those for the thin-film configuration, but in some cases they were even lower than that of Sn-based anodes, even though the theoretical capacity of $Sn_{4.4}Li$ is significantly lower than that of $Si_{4.4}Li$. For example, Si nanoparticles, 80 nm in diameter, gave a capacity of 1700 mAh/g on the tenth cycle [26], while electrochemically cycling Si nanoclusters 50 times reduced their capacity to 525 mAh/g, even though their starting capacity was 2400 mAh/g [27]. In [50], this capacity fade was again documented, showing that Si nanoparticles gave starting capacities of 2400 mAh/g, which reduced to 600 mAh/g after 20 cycles, implying a capacity retention of only 25% [50]. These poor electrochemical properties of Si nanocrystals are attributed to severe fracture that takes during cycling.

6.7.2.1 Si–SiO$_2$–C Composites

In an attempt to reduce fracture and therefore achieve a better capacity retention, Si particles were embedded in a SiO_2 matrix, anticipating that the SiO_2 would buffer the volume expansions of the Si particles. In order to make the deformation mechanisms of the Si active sites less severe, the Si nanocrystals were taken to be 2–10 nm in diameter [51]. Using solely SiO_2 as a matrix material is not efficient since SiO_2 is nonconductive and traps the Li ions within it; hence a Si–SiO_2–C composite anode was proposed [51]. To fabricate the aforementioned composite, a mixture of SiO powder milled with graphite powder was dispersed into monomer solution and then polymerized. With heat treatment, Si nanocrystals and amorphous clusters were dispersed in SiO_2, which in turn were embedded in graphite. It should be noted that Si reacts with C to form SiC above 1000 °C, so it is necessary to maintain appropriate temperatures in order to inhibit the formation of this additional nonconductive phase. Upon electrochemical cycling, Li enters the pure Si active sites and once they are fully lithiated, it is intercalated with C. Li, though, is also trapped in the SiO_x phases by forming Li_2SiO_3 and Li_2SiO_4; these phases are formed irreversibly during the first Li-insertion process, and therefore both the amount of Si and Li that can control the capacity of the battery is reduced. The

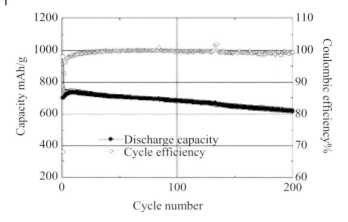

Figure 6.24 Capacity retention of Si–SiO$_2$–C nanocomposite. (Reproduced from Ref. [51].)

initial capacity of this Si–SiO$_2$–C anode was 700 mAh/g and after 200 cycles it had reduced to 620 mAh/g [51]. It was noted in [51] that the smaller the active sites, the higher the capacity retention. In Figure 6.24, it is seen that the columbic efficiency was 100% after the first few cycles. The lower coulombic efficiency observed initially is due to the Li reaction with the surface of the anode forming the SEI layer.

In order to optimize the performance of these promising anode materials, the wt.% of the oxide phase must be reduced. A step toward this direction was taken in [50], where the performance of pure Si and Si–SiO$_2$ electrodes, with 12 wt.%O, was compared directly. The pure Si nanocrystals had an initial capacity of 2400 mAh/g and a 25% capacity retention after 20 cycles, while the oxide-covered Si (Figure 6.25) not only had a lower starting capacity of 1500 mAh/g but also a lower capacity retention of 21%. This suggested that the SiO$_2$ limits the diffusivity of Li into the Si particles, and therefore C was used as the protective matrix, as opposed to SiO$_2$ [50]. The Si–C composites were fabricated by mixing Si powder with a resorcinol-formaldehye (RF) gel and then carbonizing the mixture to produce Si particles coated by hard C, as shown in Figure 6.26. The disadvantage of this technique is that Si reacted with the oxygen in the RF gel to produce SiO that covered a certain surface area of the Si particles and during the carbonization, this SiO decomposed to Si and SiO$_2$. Due to the presence of SiO$_2$, the anticipated Si–C core–shell composite gave initial capacities of 950 mAh/g, which reduced to 350 mAh/g after 20 cycles.

Although the nanocomposite of [50] resulted in lower capacities than that earlier developed in [51], it was still described here to illustrate the significance of using small particle active sites. The fact that the Si-based composite of [51] had a significantly higher capacity retention than that of [50], despite the existence of higher amounts of nonconductive SiO$_x$ phases in [51], is attributed to the fact that the nanocrystalline Si particles were significantly smaller in [51]; in [51], the Si diam-

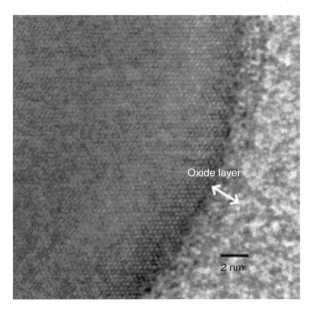

Figure 6.25 Si nanoparticle covered by a SiO layer. (Reproduced from Ref. [50].)

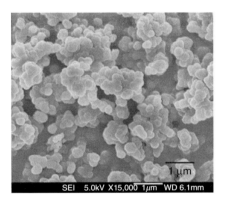 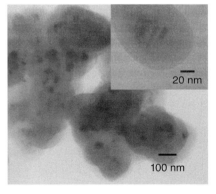

Figure 6.26 Carbon-coated Si; SEM and TEM images. (Reproduced from Ref. [50].)

eter was 2–10 nm, whereas in [50] it was 30–50 nm. These larger particles resulted in more severe cracking that disrupted the connectivity of the anode, as seen in Figure 6.27, and hence gave poor capacity retention.

6.7.2.2 Si–C Nanocomposites

Significantly higher capacities than those in which O was present have indeed been achieved in Si-sol gel graphite (Si/SGG) anodes [52]. The Si particles were dispersed in the SGG through planetary ball milling and hand grinding. The 3D

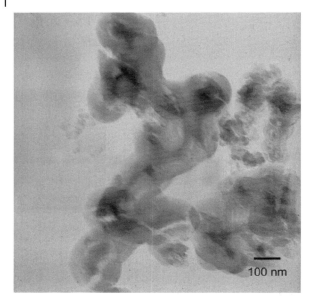

Figure 6.27 TEM image of a silicon–carbon core–shell composite after 30 cycles. (Reproduced from Ref. [50].)

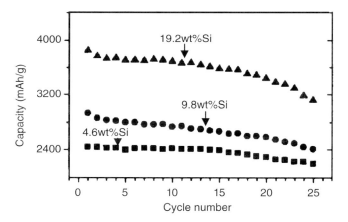

Figure 6.28 Affect of Si content in Si–SGG composites. (Reproduced from Ref. [52].)

structure of the SGG was very effective in buffering the Si volume expansions, and the capacities obtained were greater than 2400 mAh/g for 25 cycles. However, it can be seen in Figure 6.28 that with continuous cycling, there seemed to be a tendency for further capacity fade. Allowing for higher Si contents (19.8 wt.%Si) the initial capacity approached 4000 mAh/h; however, the capacity fade was more

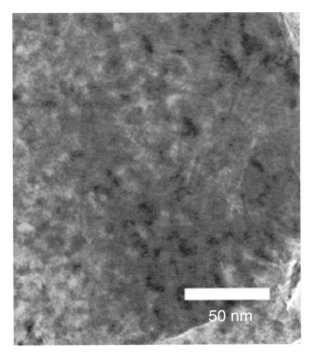

Figure 6.29 Si nanoparticles embedded in amorphous carbon. (Reproduced from Ref. [53].)

pronounced. Reducing the Si content allowed a better capacity retention, indicating that not only the Si particle size, but also the amount of Si affect the capacity. In Chapter 8, a theoretical justification, based on fracture mechanics, will justify the qualitative trend of Figure 6.28.

Another, promising Si–C composite was fabricated in [53]. 20 wt.%Si–80 wt.%C composites (Figure 6.29) were fabricated through a dehydration reaction using concentrated H_2SO_4 as the dehydration agent and thermal decomposition of sucrose. The Si particle sizes were below 25 nm, and they were embedded in C such that the metal active surfaces were protected and also agglomeration was prevented. Since the C content is significantly higher that the Si content, the initial capacity was 1000 mAh/g, and a 100% retention was obtained for 30 cycles, as seen in Figure 6.30. This unique performance is attributed to the mechanical stability that was attained through the nanosize of the Si particles and the buffering support that was offered by the carbon matrix, as well as the good conducting properties of carbon. It is seen in Figure 6.30 that embedding the Si particles in a Ag matrix resulted in a significant capacity fade, while pure Si particles provided a lower capacity than graphite after the fifth cycle.

In concluding this subsection, it is interesting to note that nano-Si/cellulose nanocomposites have been proven to be of the most promising active/inactive anodes [54a], allowing for capacities of 1400 mAh/g to be attained for 50 cycles, as

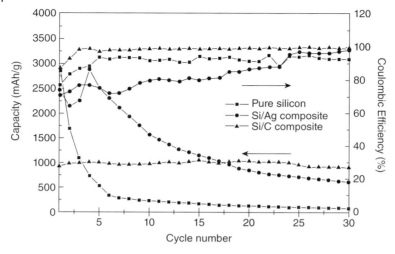

Figure 6.30 Comparison between capacity retention in pure Si, Si/Ag, and Si/C nano-composites. (Reproduced from Ref. [53].)

(a)   (b)

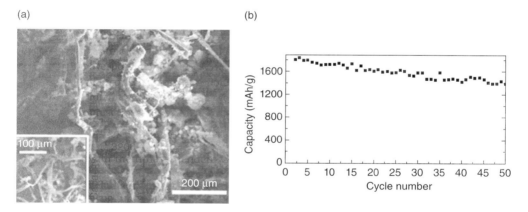

Figure 6.31 (a) SEM image of Si nanoparticles attached on cellulose fibers; inset indicates fibers prior Si adhesion. (b) Capacity retention of a 34%Si/cellulose composite, cycled at 200 mA. (Reproduced from Ref. [54a].)

seen in Figure 6.31. The structure of these most promising Si-based anodes is similar to that of the Sn/C nanocomposites described in Section 6.7.1.3, since as the Si nanoparticles (40–50 nm) were mixed with the cellulose in isopropanol and stirred, they adhered onto the cellulose fibers, thus forming a type of island structure, as seen in Figure 6.31. The high capacity obtained from these materials is attributed to the mechanical stability of the Si particles offered by the cellulose fibers. It should be mentioned that this microstructure was initially attained for Sn particles on cellulose [54b]; however, the resulting electrochemical properties were not comparable to that of the Sn-based anodes in the previous section.

6.8
Other Anode Materials

Although most the research in metal-based anodes is focused on having Sn and Si as the active sites due to their very high capacities, 996 and 4200 mAh/g, respectively, there exist other metals that are active with respect to Li as mentioned in the introduction. Although the resulting capacities from using such metal-based anodes are much lower, it is of interest to include some examples here.

6.8.1
Sb-Based Anodes

In Section 6.7.1.1, it was shown that SnSb alloys can be used as anodes. Sn is the most active material with respect to Li, and therefore such anodes are considered as Sn based; the main role of the Sb is to constrain the expansion of the Sn, and inhibit agglomeration of the Sn sites. Upon lithiation, Sb gives a capacity of 660 mAh/g, and experiences an expansion of 150%. In order to accommodate this expansion, Sb has been alloyed with Cu producing a Cu_2Sb intermetallic alloy; the structure of these alloys are alternating layers and upon Li insertion the phases produced are [55]

$$Cu_2Sb + 2Li \leftrightarrow Li_2CuSb + Cu$$

$$Li_2CuSb + Li \leftrightarrow Li_3Sb + Cu.$$

Depositing a powdered Cu_2Sb anode on a Cu foil current collector allowed for a capacity of 290 mAh/g, which was retained after 25 cycles, over the full voltage range 0–1.2 V. Although this capacity is lower than that of graphite, it is near the theoretical capacity of this material, which is 323 mAh/g [56]. In [57], it was shown that a capacity of 290 mAh/g could also be attained for Cu_2Sb thin films fabricated through heat treatments and electrodeposition.

Higher capacities have been achieved by embedding pure 100–200 nm Sb particles in a pyrolytic polyacrylonitrile polymer matrix [58]. The second discharge capacity was 485 mAh/g, while after the 20th cycle it had reduced to 408 mAh/g.

About the same time that the optimum microstructure of Figure 6.31 for Si was obtained, the potential of such a nanocomposite was also illustrated for Sb particles that are adhered on cellulose [59] as shown in Figure 6.32. The resulting capacity of the Sb/cellulose nanocomposite was 470 mAh/g for 40 cycles (Figure 6.33), while the capacity of pure Sb nanoparticles degraded below 70 mAh/g after the 15th cycle. Hence, this island-type structure seemed to be most promising for Sb as well.

In [60], another promising Sb-anode was proposed. It was shown that cycling $LiSbO_3$ against Li metal gave starting capacities that were nearly 600 mAh/g, but a significant drop to 500 mAh/g was observed at the 15th cycle. In particular, upon lithiation the reversible electrochemical reaction that takes place is

$$LiSbO_3 + 4Li^+ + 4e^- \leftrightarrow Li_5SbO_3$$

Figure 6.32 (a) Sb nanoparticles adhered on the surface of cellulose fibers, (b) unattached Sb nanoparticles. (Reproduced from Ref. [59].)

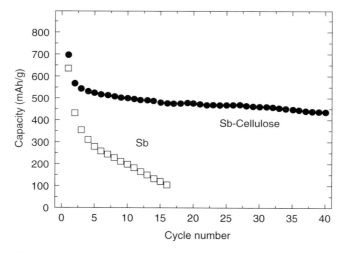

Figure 6.33 Comparison between capacity retention in pure Sb nanoparticles and Sb/cellulose nanocomposite. (Reproduced from Ref. [59].)

This suggested that other materials of the same family such as $LiSb_3O_8$ and Li_3SbO_4 could also be anode candidates.

6.8.2
Al-Based Anodes

Al–Li alloys give a wide range of theoretical capacities depending on the Li content; AlLi: 993 mAh/g (similar to that of $Li_{4.4}Sn$), Al_2Li_3: 1490 mAh/g, and Al_4Li_9: 2235 mAh/g. These high capacities are partly due to the lower atomic weight of Al, as compared to that of Sn. It is very difficult, however, to maintain such high capacities after the first cycle. This may be partly attributed to the fact that the Li diffusion coefficient in Al (6×10^{-12} cm^2/s at 293 K) [61] is two orders of magnitude lower than that in Sn [62].

Just as in most metal-based anodes, the first Al-anode configuration that was examined was the thin film form. The capacity increased as the Al thin film thickness decreased. The thinnest film examined [62] was 0.1 µm and it gave an initial capacity of 800 mAh/g; however, it was not possible to retain it.

Powder Al particles have also been fabricated again without the possibility to retain the initial capacity. Various particle sizes in the micron scale were examined, but a coherent trend relating reduction in particle size with better capacities was not obtained [63], and therefore further studies need to be performed. In [64], it was documented that pure Al particles gave starting capacities of 789 mAh/g, which dropped to 325 mAh/g after 10 cycles. This significant capacity decrease is attributed to severe cracking and pulverization that occurs from the first cycle, as seen in Figure 6.34. In order to protect the Al particles from cracking and from contact with the electrolyte, to both of which the reduction of capacity is attributed, SnO nanoparticles were attached, through chemical means on the Al surface, as seen in Figure 6.35. After one cycle, it was observed that no damage was apparent for such composite particles due to an increase in fracture strength, tensile strength, and elastic modulus of reinforced Al composites [65]. On initial Li insertion, SnO decomposes to Li$_2$O and Sn, and therefore the conductivity is also increased. The possibility of Sn-lithiation is not touched upon in [64]. In particular, three different wt.% for the SnO coating were considered: 5%SnO, 10 wt.%SnO, and 20 wt.%SnO. In Figure 6.36, it is seen that the Al sample covered with 10 wt.%SnO showed the best electrochemical properties, giving a starting capacity of 810 mAh/g, which reduced to 582 mAh/g after 10 cycles. Figure 6.36, however, suggests that further cycling will result in additional capacity reduction. Finally, it is noted that increasing the SnO content to 20 wt.% or reducing it to 5 wt.% gave

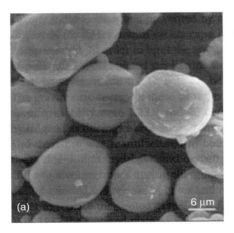

Figure 6.34 (a) Pure Al particles before cycling; (b) pure Al particles after first cycle, against a Li electrode. (Reproduced from Ref. [64].)

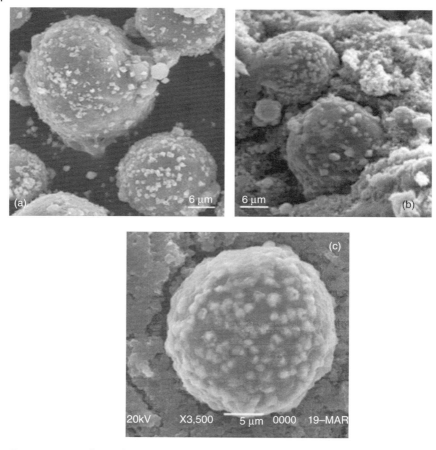

Figure 6.35 (a) Al particles covered with 10wt.%SnO nanoparticles before cycling; (b, c) Al particles covered with 10wt.%SnO after first cycle. (Reproduced from Ref. [64].)

lower capacities. This shows that the amount of SnO must be such so as to protect the Al from fracture and pulverization, but not so much so as to bind a significant amount of Li, upon the decomposition of SnO.

6.8.3
Bi-Based Anodes

The final alternative Li-active material that will be described is Bi. The theoretical capacity upon the formation of Li_3Bi is 385 mAh/g [66], which is practically near that of currently used graphite anodes, and therefore from a commercial point of view Bi is not promising. From a scientific viewpoint, however, it is of interest to look at the electrochemical cycling properties of Bi since it is consistent with previous observations on Sn- and Si-based anodes.

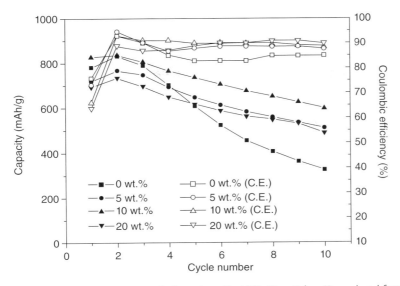

Figure 6.36 Capacity versus cycle for various Al-wt.%SnO particles. (Reproduced from Ref. [64].)

Figure 6.37 Micrographs showing Bi nanoparticles of 300 nm diameter (a), and dendrite formation after 10 cycles (b); Reproduced from Ref. [16a].

Upon maximum Li insertion, Bi expands 210%, and as a result it also has a low cyclability in the bulk and microscale due to severe deformation. In order to decrease the volume expansion damage, nanoscale Bi particles with an average diameter of 300 nm were embedded inhomogeneously in a graphite-PVDF/HFP SOLEF copolymer (Solvay) which acts as a binder. A 50% capacity fade occurred after ten cycles; the Bi content of the anode was varied to various weight%, and it was noted that lower Bi content allowed for a better capacity retention; similar observations were made for Si in Figure 6.28. This low capacity retention was attributed to the fact that the Bi nanoparticles were distributed randomly, favoring, therefore, agglomeration and dendrite formation (Figure 6.37) that resulted from the 210% volume expansion upon Li insertion.

6.9
Conclusions

In concluding this overview, it should be mentioned that the research articles published in this area, every year, are plethora and it would be impossible to summarize all of them. The purpose was to describe some of the most promising configurations and fabrication methods for the next-generation anodes. General conclusions that can be drawn are that fiber-like/nanowire configurations of active materials such as Si and Sn accommodate the colossal volume expansions and allow for better capacity retentions than the thin film case. Furthermore, it was illustrated that active/less-active nanocomposites improve significantly the electrochemical properties of all metal-based anodes, while the most promising microstructure is that of active site islands attached on the surface of carbon or cellulose (which also contains carbon). It is anticipated that within the next couple of years, some type of materials selection, most likely an Sn–C or Si–C, nanocomposite will be used commercially in the next-generation Li batteries.

References

1 Murphy, D.W. and Christian, P.A. (1979) *Science*, **205**, 651.

2 Shu, Z.X., McMillan, R.S., and Murray, J.J. (1993) *J. Electrochem. Soc.*, **140**, 922.

3 Fong, R., von Sacken, U., and Dahn, J.R. (1990) *J. Electrochem. Soc.*, **137**, 2009.

4 (a) Sato, K., Noguchi, M., Demachi, A., Oki, N., and Endo, M. (1994) *Science*, **264**, 556.
(b) Shu, Z.X., McMillan, R.S., and Murray, J.J. (1993) *J. Electrochem. Soc.*, **140**, 992.

5 Yang, J., Winter, M., and Besenhard, J.O. (1996) *Solid State Ion..*, **90**, 281.

6 Rao, B.M.L., Francis, R.W., and Christopher, H.W. (1977) *J. Electrothem. Soc.*, **124**, 1490.

7 Besenhard, J.O. (1978) *J. Electroanal. Chem.*, **94**, 81.

8 Besenhard, J.O. and Fritz, H.P. (1975) *Electrochim. Acta*, **20**, 513.

9 Wang, J., Raistrick, I.D., and Huggins, R.A. (1986) *J. Electrochem. Soc.*, **133**, 457.

10 Beaulieu, L.Y., Eberman, K.W., Turner, R.L., Krause, L.J., and Dahn, J.R. (2001) *Electrochem. Solid-State Lett.*, **4**, A137.

11 Konstantinidis, D.A. and Aifantis, E.C. (1998) *Nanostructr. Mater.*, **10**, 1111.

12 Aifantis, K.E. and Konstantinidis, A.A. (2009) *Mater. Sci. Eng. A*, **503**, 198.

13 (a) Gao, B., Sinha, S., Fleming, L., and Zhou, O. (2001) *Adv. Mater.*, **13**, 816.
(b) Wen, C.J., and Huggins, R.A. (1981) *J. Solid. State. Chem.*, **37**, 271.
(c) Wakihara, M. (1999) Extended Abstracts, A–34, 92, *12th Int. Conf. on Solid State Ionics SSI–12*, Halkidiki, Greece, June 6–12, 1999.

14 Idota, Y., Mishima, M., Miyiaki, Y., Kubota, T. and Miyasaka, T. (1997) US Patent 5, 618,641.

15 Idota, Y., Kubota, T., Matsufuji, A., Maekawa, Y., and Miyasaka, T. (1997) *Science*, **276**, 1395.

16 (a) Crosnier, O., Brousse, T., Devaux, X., Fragnaud, P., and Schleich, D.M. (2001) *J. Power Sources*, **94**, 169.
(b) Courtney, I.A. and Dahn, J.R. (1997) *J. Electrochem. Soc.*, **144**, 2045.

17 Li, H., Huang, X., and Chen, L. (1999) *J. Power Sources*, **81–82**, 340.

18 Stjerndahl, M., Bryngelsson, H., Gustafsson, T., Vaughey, J.T., Thackeray, M.M., and Edström, K. (2007) *Electrochim. Acta*, **52**, 4947.

19 Aurbach, D., Gamolsky, K., Markovsky, B., Gofer, Y., Schmidt, M., and Heider, U. (2002) *Electrochim. Acta*, **47**, 1423.

20 Chen, L., Wang, K., Xie, X., and Xie, J. (2007) *J. Power Sources*, **174**, 538.

21 Brousse, T., Retoux, R., Herterich, U., and Schleich, D.M. (1998) *J. Electrochem. Soc.*, **145**, 1.

22 Li, N., Martin, C.R., and Scrosati, B. (2001) *J. Power Sources*, **97**, 240.

23 Brousse, T., Lee, S.M., Pasquereau, L., Defives, D., and Schleich, D.M. (1998) *Solid State Ion.*, **113**, 51.

24 Li, Y., Tu, J.P., Huang, X.H., Wu, H.M., and Yuan, Y.F. (2006) *Electrochim. Acta*, **52**, 1383.

25 Li, Y., Tu, J.P., Huang, X.H., Wu, H.M., and Yuan, Y.F. (2007) *Electrochem. Commun.*, **9**, 49.

26 Li, H., Huang, X., Chen, L., Wu, Z., and Liang, Y. (1999) *Elecrochem. Solid-State Lett.*, **2**, 547.

27 Graetz, J., Ahn, C.C., Yazami, R., and Fultz, B. (2003) *Electrochem. Solid-State Lett.*, **6**, A194.

28 Maranchi, J.P., Hepp, A.F., and Kumta, P.N. (2003) *Electrochem. Solid-State Lett.*, **6**, A198.

29 Lee, K., Jung, J., Lee, S., Moo, H., and Park, J. (2004) *J. Power Sources*, **129**, 270.

30 Moon, T., Kim, C., and Park, B. (2006) *J. Power Sources*, **155**, 391.

31 Takamura, T., Ohara, S., Uehara, M., Suzuki, J., and Sekine, K. (2004) *J. Power Sources*, **129**, 96.

32 Yin, J., Wada, M., Yamamoto, K., Kitano, Y., Tanase, S., and Sakai, T. (2006) *J. Electrochem. Soc.*, **153**, A472.

33 Yang, H., Fu, P., Zhang, H., Song, Y., Zhou, Z., Wu, M., Huang, L., and Xu, G. (2007) *J. Power Sources*, **174**, 533.

34 (a) Ahn, D., Kim, C., Lee, J.-G., and Park, B. (2008) *J. Solid State Chem.*, **181**, 2139.
(b) Zeng, Z.Y., Tu, J.P., Wang, X.L., and Zhao, X.B. (2008) *J. Electroanal. Chem.*, **616**, 7.

35 Ahna, J.-H., Kim, Y.-J., and Wang, G. (2008) *J. Alloys Compd.*, doi: 10.1016/j.jallcom.2008.07.227.

36 Ishihara, T., Nakasu, M., Yoshio, M., Nishiguchi, H., and Takita, Y. (2005) *J. Power Sources*, **146**, 161.

37 Cui, L.-F., Ruffo, R., Chan, C.K., Peng, H., and Cui, Y. (2009) *Nano Lett.*, **9**, 491.

38 Wang, Z., Tian, W., and Li, X. (2007) *J. Alloys Compd.*, **439**, 350.

39 Wang, K., He, X., Ren, J., Wang, L., Jiang, C., and Wan, C. (2006) *Electrochim. Acta*, **52**, 1221.

40 Harreld, C.S. and Schlemper, E.O. (1971) *Acta Cryst.*, **B27**, 1964.

41 Seo, J.-W., Jang, J.-T., Park, S.-W., Kim, C., Park, B., and Cheon, J. (2008) *Adv. Mater.*, **20**, 4269.

42 (a) Momma, T., Shiraishi, N., Yoshizawa, A., Osaka, T., Gedankan, A., Zhu, J., and Sominski, L. (2001) *J. Power Sources*, **97–98**, 198.
(b) Mukaibo, H., Yoshizawa, A., Momma, T., and Osaka, T. (2003) *J. Power Sources*, **119**, 60.
(c) Elidrissi, M.M., Bousquet, C., Olivier-Fourcade, J., and Jumas, C. (1998) *Chem. Mater.*, **10**, 968.

43 Lee, J.Y., Zhang, R., and Liu, Z. (2000) *J. Power Sources*, **90**, 70.

44 Santos-Peña, J., Brousse, T., and Schleich, D.M. (2000) *Solid State Ion.*, **135**, 87.

45 Balan, L., Schneider, R., Ghanbaja, J., Willmann, P., and Billaud, D. (2006) *Electrochim. Acta*, **51**, 3385.

46 Sarakonsri, T., Aifantis, K.E., and Hackney, S.A. (2010) *Nanostruct. Matls.*, In Print.

47 Aifantis, K.E., Brutti, S., Sarakonsri, T., Hackney, S.A., and Scrosati, B. (2010) *Electrochim. Acta*, In Print.

48 Kim, I., Blomgren, G.E., and Kumta, P.N. (2004) *Electrochem. Solid-State Lett.*, **7**, A44.

49 Derrien, G., Hassoun, J., Panero, S., and Scrosati, B. (2007) *Adv. Mater.*, **19**, 2336.

50 Jung, Y.S., Lee, K.T., and Oh, S.M. (2007) *Electrochim. Acta*, **52**, 7061.

51 Morita, T. and Takami, N. (2006) *J. Electrochem. Soc.*, **153**, A425.

52 Niua, J. and Lee, J.Y. (2002) *Electrochem. Solid-State Lett.*, **5**, A107.

53 Wen, Z., Yang, X., and Huang, S. (2007) *J. Power Sources*, **174**, 1041.

54 (a) Gómez Cámer, J.L., Morales, J., and Sánchez, L. (2008) *Electrochem. Solid-State Lett.*, **11**, A101.
(b) Caballero, A., Morales, J., and Sánchezz, L. (2005) *Electrochem. Solid-State Lett.*, **8**, A464.

55 Thackeray, M.M., Vaughey, J.T., and Johnson, C.S. (2003) *J. Power Sources*, **113**, 124.

56 Fransson, L.M.L., Vaughey, J.T., and Benedek, R. (2001) *Electrochem. Commun.*, **3**, 317.

57 Bryngelsson, H., Eskhult, J., Nyholm, L., and Edström, K. (2008) *Electrochim. Acta*, **53**, 7226.

58 He, X., Pu, W., Wang, L., Ren, J., Jiang, C., and Wan, C. (2007) *Electrochim. Acta*, **52**, 3651.

59 Caballero, A., Morales, J., and Sánchez, L. (2008) *J. Power Sources*, **175**, 553.

60 Kundu, M., Mahanty, S., and Basu, R.N. (2009) *Electrochem. Commun.*, **11**, 1389.

61 Hamon, Y., Brousse, T., Jousse, F., Topart, P., Buvat, P., and Schleich, D.M. (2001) *J. Power Sources*, **97–98**, 185.

62 Huggins, R.A. (1999) *J. Power Sources*, **81-82**, 13.

63 Lei, X., Wang, C., Yi, Z., Liang, Y., and Sun, J. (2007) *J. Alloys Compd.*, **429**, 311.

64 Lei, X., Xiang, J., Ma, X., Wang, C., and Sun, J. (2007) *J. Power Sources*, **166**, 509.

65 Venkateswara Rao, K.T., and Ritchie, R.O. (1992) *Int. Mater. Rev.*, **37**, 153.

66 Crosnier, O., Devaux, X., Brousse, T., Fragnaud, P., and Schleich, D.M. (2001) *J. Power Sources*, **97**, 188.

7
Next-Generation Electrolytes for Li Batteries

Soo-Jin Park, Min-Kang Seo, and Seok Kim

7.1
Introduction

At the turn of the millennium, the consciousness of the need to use energy more efficiently spread, at least in the "developed" countries. The efficient use of energy also includes efficient storage of electricity, meaning that one needs to have high energy density batteries that have low energy losses during storage, electrical charging and discharging, and last, but not the least, batteries with long lifetimes and minimum production and disposal costs [1, 2].

In the last decade, a number of different battery systems were found to meet the requirements for electric vehicles (EV), among which are the nickel–metal hydride, the sodium–sulfur, and the lithium-ion systems [3]. In the USA, the lithium-ion battery is considered to be most promising and thus its development is considerably supported [2]. The history of the development of secondary lithium batteries was described in Chapter 3, but additional information can be found in [4].

There are two major arguments for using a lithium-ion battery: lithium is the lightest element that can safely be handled in electrochemical processes and it exhibits the highest oxidation potential of any element; properties that make lithium ideally suited for high energy density batteries. However, lithium–metal electrodes in contact with liquid electrolytes cause a variety of problems, which in the worst case can lead to fires and explosions. Therefore, the most promising approach is that of the "rocking chair" cell, in which the lithium metal is replaced by a lithium-ion source [5–7]; examples of such, current and future, electrodes are given in Chapters 5 and 6. During charging and discharging of such batteries, lithium ions are rocked between lithium–carbon and lithium–metal oxide intercalation compounds, which act as the electrode couple. A typical example is displayed in Figure 7.1, in which Li_xC_6 is used as the anode and $Li_xMn_2O_4$ (a spinel) as the cathode.

Due to the numerous applications of lithium batteries in the automotive and aerospace industries and portable devices, significant interest exists from both developers and manufacturers on improving their material chemistries [1, 8, 9].

High Energy Density Lithium Batteries. Edited by Katerina E. Aifantis, Stephen A. Hackney, and R. Vasant Kumar
© 2010 WILEY-VCH Verlag GmbH & Co. KGaA, Weinheim
ISBN: 978-3-527-32407-1

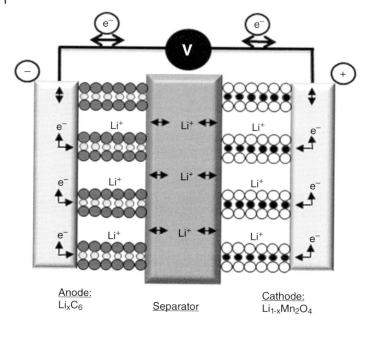

anode reaction: $Li_xC_6 \Longleftrightarrow xLi^+ + xe^- + C_6$

cathode reaction: $xLi^+ + xe^- + Li_{1-x}Mn_2O_4 \Longleftrightarrow LiMn_2O_4$

charging $\Longleftrightarrow$ discharging

Figure 7.1 Schematic illustration of a lithium rocking chair battery with graphite and spinel as intercalation electrodes and its electrode reactions.

As cathode and anode materials selection was described in the previous chapters, the present chapter will focus on alternative electrolytes. Current liquid electrolytes employed in Li-ion cells result in a gradually degradable charge–discharge cycle performance, due to volatility problems [10, 11], and decomposition reactions of the liquid organic electrolyte solution (e.g. EC:PC, DMC-containing $LiPF_6$) that may result from the high oxidation potential of the metal oxide cathodes. Therefore, solid or solid-like electrolytes are in demand for use in high-capacity Li batteries [12–15]. So far it has been possible to commercialize Li batteries using polymer gel electrolytes (PGE), not solid polymer electrolytes (PE). AU Li batteries, however, that employ a form of polymer as their electrolyte are commercially termed Li polymer batteries.

At present, various research units and private companies such as PolyPlus, Moltec, and Ultralife Batteries in the USA and Maxell in Japan are actively involved in developing polymer lithium batteries both for small- and large-scale applications. The PolyPlus Battery in the US, for example, is developing room temperature lithium polymer batteries, which can deliver a high specific energy (200 Wh/kg) and a high rate discharge (500 W/kg). In general, the solid redox polymerization electrode (SRPE) consists of the electroactive materials (organodisulfide polymer) and carbon powder, in a matrix of a suitable ionically conducting polymer such as polyethylene oxide (PEO). The polydisulfide undergoes scission to form dithiolate salts upon reduction, and is reoxidized reversibly to the polymeric disulfide. Therefore, in a prototype cell working at 80–90 °C and using a lithium intercalated disulfide polymer as the cathode, and carbon as the anode (just a in Li-ion cells) a specific energy as high as 200 Wh/kg was observed [16, 17]. The discharge/charge cycles were almost reproducible for over 350 cycles. Moltec, on the other hand, has reported an AA-sized battery based on an organosulfur cathode with a specific energy density of 180 Wh/kg [18]. Ultralife Batteries have reported a solid polymer battery based on intercalation type electrodes that operates at room temperature giving 170 Wh/kg, and showing a performance of more than 500 charge/discharge cycles [19].

One of the main purposes to develop polymer lithium batteries is for traction applications. Based on the survey on demand for alternative fuel vehicles made by Turrentine and Kurani in the US in 1995 [20], consumers agreed to buy EVs, which would run for at least 200 km per battery. This was accepted mainly on the grounds of environmentally friendliness of an EV. It is only recently that details of EVs, in the US are communicated to the people through mail, telephone, and Internet communications [21]. It was found that some people rejected the idea of an EV mainly because of the need of home recharging and shorter driving range per battery. Nevertheless, the concept of zero emission and high reliabilities of an EV would still prove encouraging for an environmentally conscious person. Table 7.1 indicates the USABC (The United States Advanced Battery Consortium) criteria for an EV battery.

Table 7.1 The United State advanced battery consortium (USABC) criteria for EV battery.

Battery properties	Unit	Mid term (~2000)	Long term (~2010)
Power density	W/l	250	600
Specific power	W/kg	150–200	400
Energy density	Wh/l	135	300
Specific density	Wh/kg	80–100	200
Life	Years	5	10
	Cycles	600	1000
Price	$/kWh	<150	<100
Normal recharge time	H	<6	3–6
Operating environment	°C	−30–65	−40–85

Solid and solid-like polymer electrolytes offer unique advantages such as satisfactory mechanical properties, ease of fabrication as thin films and formation of proper electrode–electrolyte interfaces. Polymeric electrolytes have found wide application in electric or medical devices, electroluminescence devices, and photoelectric devices. Since Wright *et al.* discovered in 1973 that PE based on PEO $(OCH_2CH_2)_n)$ with an alkaline metal salt offered ionic conductivity [22], PEs have received widespread attention. Moreover, it is possible to substitute liquid electrolytes in transitional lithium-ion batteries, thereby achieving an electrolyte system for solid-like lithium polymer batteries [23–26]; therefore, the promising next-generation anodes, described in Chapter 5, could be used in Li polymer cells as well.

The field of PE has gone through three stages: "dry solid systems," "polymer gels," and "polymer composites." The "dry systems" use the polymer host as the solid solvent and do not include any organic liquids, such as $PEO/LiClO_4$. The "polymer gels" contain organic liquids as plasticizers, which with a lithium salt remain encapsulated in a polymer matrix (e.g., $PMMA/EC-PC/LiClO_4$), whereas the "polymer composites" include high surface area inorganic solids in proportion with a "dry solid polymer" or a "polymer gel" system, such as $PEO-LiClO_4$+inorganic fillers. More detailed explanations of the "dry solid systems" are excluded in this work due to the insufficient experimental data of the systems.

To compare the performance of liquid electrolytes, polymer electrolytes, and polymer composite eletrolytes, we considered three electrolytes: 1) 1 M $LiClO_4$–PC, 2) TPU–PAN–1M $LiClO_4$–PC, 3) TPU–PAN–1M $LiClO_4$–PC containing Al_2O_3 respectively (where TPU–PAN = thermoplastic polyurethane-polyacrylonitrile, and PC = propylene carbonate). Lithium metal and $LiCoO_2$ were employed as negative and positive electrodes, respectively. The area of both electrodes was fixed as $2.25\,cm^2$. $Li/PE/LiCoO_2$ laminated cells were assembled by pressing Li, electrolyte, and $LiCoO_2$; sealed by a polyethylene film, and laminated by an aluminum foil. Afterward, the cells were cycled with the BT-2043 system (Anbin electrochemical instrument, USA) between 4.2 and 2.7 V at room temperature. The charge–discharge curves of the cells at the third potential cycle are shown in Figure 7.2. It is obvious that the cell at $C/20$ rate achieves a capacity of 115 mAh/g. A comparison among the discharge curves reveals that the capacity of both PEs is slightly lower than that of liquid electrolyte. The reduced capacity is attributed to the lower diffusion rate of lithium ions in the PEs as compared with that in the liquid electrolyte [27]. Significant research is, therefore, being performed to fabricate optimum PEs.

Competitor electrolytes, to PEs, that can realize solid Li batteries, are organic salts with melting points below room temperature, so-called room-temperature molten salts (RTMS), that have been the main focus of many recent scientific investigations, due to their unique physical and chemical properties [28–30]. They are nonvolatile, nonflammable, miscible with a number of organic solvents, and have a high thermal stability. From an electrochemical point of view, they offer a high ionic conductivity and a wide potential window. For these reasons, these salts have attracted much attention for their potential application to electrochemical capacitors [31, 32] or nonaqueous Li-ion batteries [33, 34].

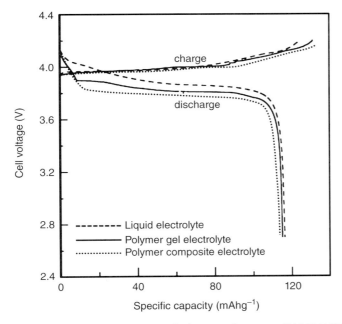

Figure 7.2 The third cycle charge–discharge performance of Li/1 M LiClO$_4$–PC/LiCoO$_2$, Li/PE (TPU–PAN-1 M LiClO$_4$–PC)/LiCoO$_2$, and Li/PE(TPU–PAN-1 M LiClO$_4$–PC Al$_2$O$_3$)/LiCoO$_2$ cells. (PE: 46.5% TPU-PAN and 54.5% LiClO$_4$–PC; 8 wt.% Al$_2$O$_3$).

The most common and extensively studied PE is based on PEO–lithium salt complexes (named PGE). This PGE shows good electrochemical and mechanical performance, and many studies on PEO combined with various lithium salts, such as LiClO$_4$ and LiBF$_4$, have been reported [35, 36]. However, PGEs have a conductivity of only 10^{-4}–10^{-5} S/cm, which is low for some applications. The unsatisfactory conductivity can be ascribed in part to the high degree of PEO crystallinity at room temperature, as well as the strong interactions between Li ions and polymers [36]. One of the efficient ways to decrease the crystallinity of a PE based on the polymer matrix, PEO, is the addition of nanoscale inorganic fillers such as TiO$_2$, SiO$_2$, and Al$_2$O$_3$, or organic plasticizers such as ethylene carbonate (EC) (C$_5$H$_6$O$_3$) or, dimethylfomamide (DMF; (CH$_3$)$_2$NC(O)H)) [14, 37–45]. These inert fillers are added to increase the volume fraction of the conductive amorphous phase. The addition to polymer films of inert fillers such as β-alumina, γ-LiAlO$_2$, etc. [41, 45] has also been pursed, to improve the ion conductivity, and the electrochemical, or mechanical properties. Notably, PEs-containing nanoparticles (named polymer composite electrolytes (PCE or CE)) exhibit better mechanical strength, higher ion conductivity, and better interfacial stability with the lithium anode. Furthermore, it has been reported that the interface between lithium and (CEs) is more stable and efficient during cycling in comparison with the filler-free electrolytes [14].

The recent interest in PCEs containing layer-structured nanoparticles, such as montmorillonite (MMT; $(Na,Ca)_{0.33}(Al,Mg)_2(Si_4O_{10})(OH)_2 \cdot nH_2O$)) and Mobil crystalline material-41, MCM-41 (MCM is the initial name given for a series of mesoporous materials that were first synthesized by Mobil's researchers in 1992; MCM-41 is Mobil Composition of Matter No. 41), has risen from the fact that such PCEs exhibit a dramatic increase in tensile strength, heat resistance, and solvent resistance, as well as, a decrease in gas permeability. All these properties are desirable in electrolytes for rechargeable batteries [46–48] and are attributed to the addition of nanoparticles.

In the present study, various additives such as 1-ethyl-3-methylimidazolium tetrafluoroborate (EMIBF$_4$), organically modified montmorillonite (OMMT), Li-exchanged MMT (Li-MMT), and mesoporous silicate (MCM-41) are, respectively, introduced into conventional PEO/LiX CEs. In this view, the effects of additive content on the microstructure and electrochemical properties of the PE are studied. The ionic conductivity at ambient and high temperatures is also analyzed. Furthermore, the electrochemical properties, such as the ion transference number and the electrochemical voltage window limit for PE stability, are investigated.

7.2
Background

7.2.1
Li-Ion Liquid Electrolytes

Besides the electrodes, the electrolyte, which commonly refers to a solution comprising of salts and solvents, constitutes the third key component of a battery. Although the role of the electrolyte is often considered trivial, its choice is actually crucial, and is based on the criteria that differ depending on whether we are dealing with polymer or liquid-based Li-ion rechargeable batteries. For instance, working with a highly oxidizing (>4 V vs Li/Li$^+$) positive electrode material for Li-ion batteries requires electrolyte combinations that operate well outside their window of thermodynamic stability (3.5 V). This is the reason why early workers in the field ignored very positive cathode materials. But fortunately this electrolyte stability is kinetically controlled, enabling the use of nonaqueous electrolytes at potentials as high as 5.5 V. Similarly, the use of a polymer rather than a liquid electrolyte adds further selection criteria linked to the electrochemical stability of the polymer. There are numerous liquid solvents available, each with different dielectric constants and viscosity, and we can select specific solvents to favor the ionic conductivity of the electrolyte. In contrast, there are only a few Li-based salts or polymers to choose from, the most commonly used ones being based on PEO. The results from research efforts aimed at counterbalancing this deficit have led to the present level of research and development on electrolytes.

Guided by general concepts of viscosity and dielectric constants, optimizing the ionic conductivity of a liquid electrolyte is almost like a field-trial approach with the hope of finding the key ingredients. For instance, only ethylene carbonate ($C_3H_4O_3$) can provide the *ad hoc* protective layer on the surface of graphite anodes that prevents further (continuous electrolyte reduction and self-discharge). Ethylene carbonate is, therefore, present in almost all commercial compositions of Li-ion cells, thinned with other solvents owing to its high melting point. Why the homologous propylene carbonate (PC; $C_4H_6O_3$) is unsuitable for this protective layer remains an open question, reminding us that chemistry has its secrets.

The present arsenal of solution components that can be produced at high enough purity and reasonable prices includes three families of solvents – namely, ethers, esters, and alkyl carbonates – and salts. Among them, alkyl carbonate solvents are most widely used in Li-ion cells.

The key features of electrolyte solutions for battery application are as follows:

1) Evaluation of transport properties – what influences conductivity and what parameters should be measured in order to evaluate the solvents properly.
2) The electrochemical stability, that is, the electrochemical window.
3) Temperature range of operation.
4) Safety features.

In general, high solvent polarity usually goes together with strong solvent–solute interactions. This not only means good solubility, but also high viscosity and high friction for ionic mobility. Hence, mixtures of solvents of high polarity and high viscosity with solvents of low polarity and low viscosity may provide optimal conductivity of Li salts (e.g., alkyl carbonates plus ethers or esters). In this respect, the salt concentration should also be optimized, since too high a salt concentration means a high concentration of charge carriers, but also strong solvent–solute and solute–solute interactions that may be detrimental to high conductivity.

The study of the electrochemical windows of the Li-ion battery electrolyte solutions has concluded that

1) The order of oxidation potentials is alkyl carbonates > esters > ethers. The oxidation of the solvents is usually the limiting anodic reaction.

2) In fact, even solvents of apparent relatively high anodic stability, such as alkyl carbonates, undergo slow scale anodic reactions on noble metal (Au, Pt) electrodes at potentials below 4 V versus Li/Li$^+$. Nevertheless, these solvents are stable with 4 V cathodes (LiNiO$_2$ LiCoO$_2$, LiMn$_2$O$_4$, etc.), whose charging potentials may reach 4.5 V (Li/Li$^+$), due to passivation phenomena. The commonly used 4 V cathodes react with solution species and become covered by surface films. These surface films seem to inhibit massive solvent oxidation at potentials below 4.5 V. Consequently, alkyl carbonate solutions may be stable with cathode materials at potentials as high as 5 V. Any negative electrode that operates at potentials below 1.5 V (Li/Li$^+$) should react with the solution species present in the electrolyte and become covered by a surface film, termed solid electrolyte interface (SEI; see Section 6.2) that is comprised of insoluble Li salts. Hence, Li or Li–C electrodes are obviously covered by

surface films and should be defined as SEI electrodes. Therefore, negative electrodes that operate at potentials >1.5 V (e.g., Li_xTiO_y compounds) may not be controlled by surface films. This SEI layer is protective of the electrode, but at the same time, if it becomes too thick it inhibits Li-ion diffusion.

3) A high oxidation potential requires a high oxidation state of the solvents' atoms and is good for the cathodes, for example, alkyl carbonate solvent liquid electrolytes. However, the high oxidation state of the solvent atoms means high reactivity at the negative side. As a result, with highly reactive anodes, such as Li metal-based systems, low oxidation state solvents/systems (e.g., ethers, PEO derivatives) should be used. This means a penalty on the positive side: with electrolytes containing ethers/PEO derivatives one cannot use 4 V cathodes. Thus, with Li metal-based systems, cathodes, such as V_2O_5, MV_2O_5 (M = Na, Ca), 3 V Li_xMnO_2, should be used, which are compatible with ethers and PEO derivatives.

The major challenges in R&D of improved electrolyte solutions for Li-ion batteries are to obtain high anodic stability for 5 V systems to improve low temperature conductivity and high temperature performance with minimal capacity-fading, and to improve safety features, meaning no thermal runaway reactions and no flammability. The key behavior factor is the electrodes' surface chemistry: good passivation at a wide temperature range. In the case of electrolytes, there are suggestions for new salts, for example, a combination of $LiPF_6$ + $LiPF_3$ $(CF_2CF_3)_3$ is promising (surface and solution bulk effect), as well as the new LiBOB salt (it is solid state but soluble in organic solvent).

The easiest, cheapest, and most effective route for the improvement of the currently used electrolyte solutions is the use of surface-active additives. However, it is important to carry out reliable tests for new solutions. Tests at elevated temperatures and measurements of self-discharge currents during storage are believed by the authors to be a suitable basis for the selection of new additives that improve the electrode/solution interactions.

7.2.2
Why Polymer Electrolytes?

A variety of reasons have been advanced for using polymer electrolytes in lithium batteries. These include increased safety, higher specific energy and energy density, greater flexibility in packaging, and lower cost.

The argument to support increased safety is more historical than real. The early problems with rechargeable cells using pure lithium metal anodes lead to a considerable amount of weight that resulted from coating the metal electrode with a polymer to protect it from rapid reaction. The introduction of the carbon anodes, however, has reduced the safety concern related to the build-up of very reactive lithium metal deposits during cycling. Nonetheless, a PE does provide reduced probability of shorting and less combustible materials compared to the liquid electrolyte cell.

The reason for higher specific energy and energy density is less mass and less volume; however, this can only be realized if the polymer electrolyte is thinner than the currently used separators in liquid electrolyte cells. This means that the polymer must have a thickness below 50 µm. Only with polymer electrolytes formed *in situ* have practical procedures been demonstrated for building cells with very thin electrolytes.

All commercial rechargeable batteries and most primary batteries are marketed in rigid packages. A flexible, prismatic package offers considerable, and potentially important, opportunities. This is the promise of polymer electrolytes; however, yet to be demonstrated are long-lived multilayer flexible packages offering a wide variety of designs. One possibility of the "dry" polymer electrolytes is practical bipolar constructions.

A feature of the polymer electrolyte technology is laminar construction. This introduces the concept of cell construction by thin film or coating processes adopted from technologies such as those found in the paper industry. The vision is of large scale, automated equipment able to carry out continuously the operations of laminar construction, cut, and package in high volume. Such processing appears to provide major cost savings over present cell and battery assembly operations.

7.2.3
Metal Ion Salts for Polymer Electrolytes

The metal ion salts can be used in electrolytes for fabricating electrochemical devices, including primary and secondary batteries, photoelectrochemical cells, and electrochromic displays. The salts have a low energy of dissociation and may be dissolved in a suitable polymer to produce a polymer solid electrolyte or in a polar aprotic liquid solvent to produce a liquid electrolyte. The anion of the salts may be covalently attached to polymer backbones to produce polymer solid electrolytes with exclusive cation conductivity.

The ion conductivity of electrolytes is related to the ability of the anion and the cation to dissociate. A low level of ionic dissociation leads to extensive ion pairing and ion clustering and lower conductivity. This effect is most pronounced in polymer electrolytes, because polymers have lower dielectric constants and a lower degree of ion complexation than polar aprotic liquid solvents typically used to produce liquid organic electrolytes. In addition to facile ionic dissociation, the electrolytes must have a high degree of thermal, chemical, and electrochemical stability.

Lithium salts that have been used to produce electrolytes for electrochemical devices have generally been selected from $LiClO$, $LiBF$, $LiAsF_6$, $LiPF_6$, and $LiSO_3CF_3$. Many of these salts are unstable or produce polymer electrolytes with relatively low conductivity. Therefore, a new class of lithium imide salts with structure $LiN(SO_2CF_3)_2$ has been introduced recently. The delocalized anionic charge facilitates dissociation of the ion pair leading to a high ionic conductivity both in liquid and polymer solid media. A methide analog lithium salt with the composition

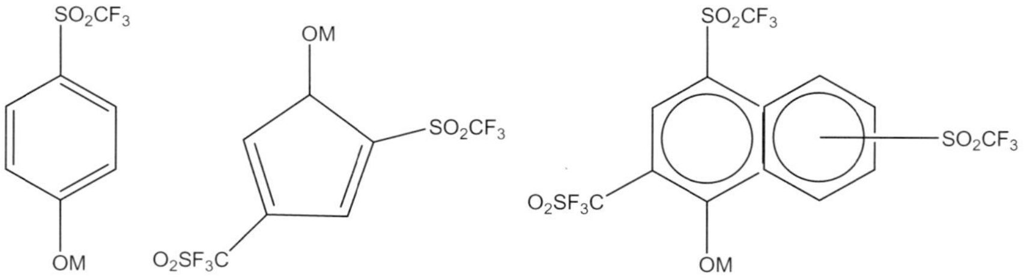

Figure 7.3 Metal ion salts with anion structure (M: Li, K, Na).

$LiC(SO–CF_3)_3$ with similar properties to the imide salts has also been introduced [49].

None of the above described salts allow covalent attachment to polymer backbones to produce polymeric ion conductors with exclusive cationic conductivity. Exclusive cationic conductivity is advantageous for electrolytes in electrochemical devices, such as batteries, as the deleterious opposing voltages produced by the counter moving anionic charges are thereby eliminated. This leads to cells that operate at higher currents and have a higher power. In contrast to the imide and methide salts described above, the lithium salts allow covalent attachment to polymer backbones.

It is apparent that in applications using polymer solid electrolytes, for example, a secondary lithium battery, it would be preferable to have no anion migration. Counter moving anions lead to polarization of the cell and a reduced power output. Accordingly, it is a primary object of the present chapter to provide a class of lithium salts, which allow covalent attachment of the anions to polymer backbones to produce polymer electrolytes with exclusive cation conductivity.

The metal ion salts are based on mononuciear or condensed aromatic moieties that have covalently attached to them one or more electron withdrawing groups of the formula SO_2CF_3 and one or more hydroxy, amino, or imino groups capable of salt formation and conjugated with SO_2CF_3 substituent through the aromatic π-electron system. The anionic charge of such conjugated ionized hydroxy, amino, or imino groups is delocalized over the whole π-electronic structure of the aromatic moiety. This delocalization effect is the key to the low lattice energy of the salts and the facile dissociation of the cation from the anion. In particular, the delocalized nature of the anionic charge results in facile dissociation of lithium salts resulting in high ionic conductivity. The chemical structures of several representative examples of metal ion salts are shown in Figure 7.3, where M is a metal ion, such as Li, Na, K.

7.3
Preparation and Characterization of Polymer Electrolytes

Now that the importance of polymer electrolytes has been described, various PE candidates will be described. First their fabrication procedure will be given and then their materials characterization.

7.3.1
Preparation of Polymer Electrolytes

7.3.1.1 Molten-Salt-Containing Polymer Gel Electrolytes

The polymer gel electrolytes (PGEs) are prepared using PEO (M_w: 2.0×10^5, manufactured by Aldrich), EC (Aldrich), and 1 mol of $LiBF_4$ with a weight ratio of PEO:EC of 6:1 with the addition of varying weight contents of $EMIBF_4$ (C-TRI Co. Ltd., Korea). The PEO is dissolved in acrylonitrile (AN, Junsei Chemical, Japan) followed by the addition of appropriate amounts of EC, $LiBF_4$, and $EMIBF_4$ at 2 h intervals and stirring for 24 h at 50 °C. The heterogeneous mixtures are poured onto glass plates and evaporated slowly at 40 °C in a vacuum oven.

7.3.1.2 Organic-Modified MMT-Containing Polymer Composite Electrolytes

In the preparation of the PCE, PEO (M_w: 2.0×10^5, Aldrich), EC (Aldrich), $LiClO_4$ (Aldrich Co. Ltd.), and montmorillonite (MMT) as inorganic fillers are used. MMT can be described as $M_x(Al_{4-x}Mg_x)Si_8O_{20}(OH)_4$, where M is the monovalent cation and x is the degree of substitution. Na^+-MMT-type layered silicate clays are available as micron size tactoids, which consist of several hundred individual particulates. They are held together by an electrostatic force with a gap of ~0.3 nm between two adjacent particles. The Na-MMT and organically modified MMT (named as OMMT) (Cloisite 20A, 25A, 30B or MMT-20A, MMT-25A, MMT-30B) supplied by Southern Clay Products (USA), are used as fillers without purification.

The PEO and $LiClO_4$ are dried in a vacuum oven for 24 h at 50 °C and 120 °C, respectively. The preparation of the PCE films first involved the dissolution of the PEO in acetonitrile and then the addition of EC. The solution is agitated for 3 h at 40 °C to completely mix the PEO and EC. The weight ratio of EC to PEO was fixed at 1:4. Then, 1 M of $LiClO_4$ is added to the solution. After 1 h agitation, 10 wt.% Na-MMT or various MMTs are added to the mixtures, after which another agitation is sustained for 24 h at 50 °C. The mixtures are then poured into Teflon plates and evaporated slowly at 40 °C in a vacuum oven to yield the PCE films.

7.3.1.3 Ion-Exchanged Li-MMT-Containing Polymer Composite Electrolytes

The ion-exchanged Li-MMT is prepared by replacing the sodium cations of the MMT clays with lithium cations. The cation-exchange capacity of the sodium montmorillonite (Na-MMT) was 1.54 meq/g. Two grams of Na-MMT were dispersed in 1 M of lithium chloride solution. The solution was stirred for 2 days to replace the sodium with lithium cations and then centrifuged to separate the Li-MMT from the supernatant containing the excess ions. The mixtures were stirred vigorously for 8 h, centrifuged, and then washed with deionized water until no chloride ions were detected with 0.1 M of $AgNO_3$ solution. The modified Li-MMT was dried in a vacuum oven at 60 °C for 24 h.

The PCE was prepared using PEO (M_w: 2.0×10^5, Aldrich), EC (Aldrich), and 0.5 M $LiClO_4$ in a weight percentage (PEO:EC = 2:1) with different weight contents of Li-MMT. The PEO was dissolved in acrylonitrile followed by the addition of appropriate amounts of EC and Li-MMT at 2 h intervals and stirring for 24 h at

50 °C. The heterogeneous mixtures were poured onto glass plates and evaporated slowly at 40 °C in a vacuum oven.

7.3.1.4 Mesoporous Silicate (MCM-41)-Containing Polymer Composite Electrolytes

MCM-41, a kind of mesoporous silicate, was synthesized following the procedure of Ryoo and Kim [50]. A sodium silicate solution with an Na/Si ratio of 0.5 was prepared by combining 46.9 g of 1.0 M NaOH solution with 14.3 g of a colloidal silica, Ludox HS40 (Du Pont), and heating the resulting gel mixtures while stirring for 2 h at 353 K. This solution was dropwise added to an aqueous NH_3 solution and 20.0 g of 25 wt.% hexadecyltrimethylammonium chloride (HTACl) solution (Aldrich). The mixtures were heated to 370 K for 24 h and then cooled to room temperature. The pH of the reaction mixtures was adjusted to 10.2 by the addition of 30 wt.% acetic acid. The mixtures were heated again to 370 K for 24 h. The above pH adjustment and heating were repeated two more times. The precipitated product was filtered and washed with distilled water and dried in an oven at 370 K. The product was calcined in air using a muffle furnace to remove the HTACl template. The temperature was increased from room temperature to 770 K over 10 h and maintained at 770 K for 4 h. The final product yield based on silicate recovery was 88%.

PCE was prepared using PEO (M_w: 2.0×10^5, Aldrich) and $LiClO_4$ (Aldrich) (mixing ratio: $P(EO)_{16}-LiClO_4$) with different weight contents of MCM-41. The PEO was dissolved in acrylonitrile followed by the addition of appropriate amounts of $LiClO_4$ and MCM-41 at 2 h intervals and stirred for 24 h at 50 °C. The mixture solution was poured onto glass plates and evaporated slowly at 40 °C in a vacuum oven.

The microstructure and properties of each of the aforementioned PEs will be described in detail in the separate subsections that follow.

7.3.2
Characterization of Molten-Salt-Containing Polymer Gel Electrolytes

7.3.2.1 Morphologies and Structural Properties

The morphology of the PEO films was studied by scanning electron microscopy, SEM, (JEOL JSM-840A SEM/LINK system AN-10000/85 S Energy Dispersive X-Ray Spectrometer) and transition electron microscopy, TEM, (A Philips CM-20) measurements. The accelerating voltage of the TEM was 160 kV.

Figure 7.4 shows SEM images of the PGE films containing various $EMIBF_4$ contents. In the case of the (a) sample, that is, a $PEO/EC/LiBF_4$ film containing no $EMIBF_4$, a rather homogeneous morphology is shown, except for some pores and some white spots probably related to the existence of salt species. By increasing the $EMIBF_4$ content from 0.1 to 0.3 mol, the film morphology was changed. First of all, some of the pores, which could have been defects in the electrolyte film, disappeared in the cases of (c) and (d). This suggested that the film compactness had been improved and that the defects had been removed by adding the $EMIBF_4$ molten salts. Besides, the white spots and lines were made more promi-

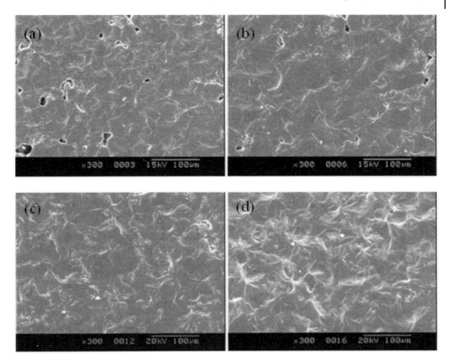

Figure 7.4 SEM images of PGE containing (a) 0, (b) 0.1, (c) 0.2, and (d) 0.3 mol of EMIBF$_4$.

nent by increasing the EMIBF$_4$ content. This implied that the additional molten salt introduced into the PGE films was enough so as to induce the precipitation of the Li salt and molten salt rich phases, resulting in the rather heterogeneous film morphologies.

The X-ray diffraction (XRD) patterns of the PE as functions of the additive concentration were obtained with a Rigaku Model D/MAX-III B diffraction-meter equipped with a rotation anode and using CuK_α radiation. The interlayer spacing of the PE containing the additives was determined by the XRD peak, using the Bragg equation

$$\lambda = 2d\sin\theta \tag{7.1}$$

where λ corresponds to the wavelength of the X-ray radiation (1.5405 Å), d corresponds to the interlayer spacing between the diffractional lattice planes, and θ is the measured diffraction angle.

Figure 7.5 displays the crystallinity changes in the electrolytes of varying molten salt contents obtained by XRD. The characteristic diffraction peaks of the PEO crystalline phase are apparent between $2\theta = 15°$ and $30°$ [51, 52]. In comparison to the pristine PEO sample, the diffraction peaks of the PGE containing EC and LiBF$_4$ became considerably smaller and less prominent, reflecting the decreased

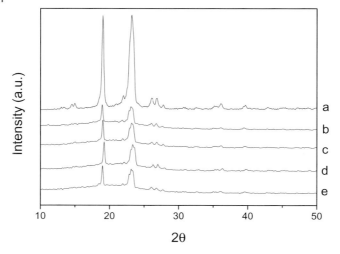

Figure 7.5 X-ray diffraction patterns of (a) pristine PEO and PGE containing (b) 0, (c) 0.1, (d) 0.2, and (e) 0.3 mol of EMIBF$_4$.

crystallinity. With the gradual addition of the additional salt to the PEO/EC system, the intensities of the characteristic peaks proportionally decreased, reaching the lowest value at a content of 3 mol. The change in the crystalline structure was also studied by means of the thermal characterization of Differential scanning calorimetry (DSC).

7.3.2.2 Thermal Properties

Differential scanning calorimetry studies were performed in the 30–125 °C range with a Perkin Elmer (DSC 6 series) system, at a heating rate of 5 °C/min and with purging nitrogen gas. Thermogravimetric analysess (TGA) were performed on the PCE using a Du Pont TGA-2950 analyzer to investigate the thermal degradation behaviors from 30 °C to 600 °C at a heating rate of 10 °C/min in a nitrogen atmosphere.

Figure 7.6 shows the DSC results for the pristine PEO and PEO/EC containing varying concentrations of additional salt. The endothermic peaks at the melting temperature (T_m), attributed to the melting of the PEO-rich crystalline phase, weakened and shifted to lower temperatures with an increase in the EMIBF$_4$ content to 0.2 mol. The PGE showed endothermal peaks due to the melting of the crystalline domain with increasing temperature. The curves moved to a low-temperature region as a result of the addition of EMIBF$_4$. It was found that the melting temperature of crystalline PEO was around 51–56 °C, which was slightly degraded when the content of EMIBF$_4$ was increased. Also, the PGE crystallinity was decreased with the addition of increasing amounts of EMIBF$_4$. Consequently, it could be stated that the presence of EMIBF$_4$ improved the ionic conductivity of the PGE by slightly decreasing the PGE crystallinity [53, 54].

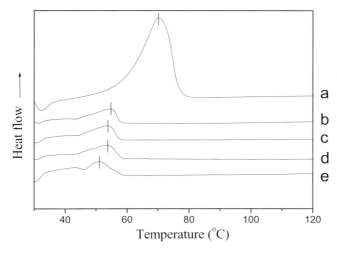

Figure 7.6 DSC curves of (a) pristine PEO and PGE containing (b) 0, (c) 0.1, (d) 0.2, and (e) 0.3 mol of EMIBF$_4$.

Table 7.2 DSC data for PGE containing various EMIBF$_4$ concentrations.

EMIBF$_4$ concentration (mol)	Melting temp., T_m (°C)	Heat of fusion, ΔH_f (J/g)	Crystallinity, χ (%)
Pure PEO	70.1	189.9	–
0	55.1	61.2	32.2
0.1	54.1	56.3	29.6
0.2	53.4	53.1	27.9
0.3	51.1	51.5	27.1

Both the XRD and DSC studies indicated that the PEO crystallinity was diminished by EC, and diminished slightly further by the addition of EMIBF$_4$. The melting temperature, heat of fusion, and relative crystallinity values are listed in Table 7.2. The PEO itself shows a heat of fusion of 189.9 J/g, and the PEO/EC (6:1) without a filler gives a heat of fusion of 61.2 J/g, the estimated crystallinity being 32.2%. This indicates that EC can reduce the crystallinity of PEO by means of plasticizer effects. With the increase in the filler content to 0.3 mol, the T_m gradually fell to 51.1 °C, and the crystallinity fell to the smallest value, 27.1%.

Figure 7.7 shows TGA graphs for PGE as functions of additional salt content. Graph (a) shows the thermogravimetric data for the pristine PEO/EC/1 mol of LiBF$_4$ film. Up to 320 °C, there was no distinct weight reduction, indicating good thermal film stability in this temperature range. However, there is an inflection point at ~300 °C, reflecting the degradation of the polymer main chain. The weight loss decelerates from 320 °C to 350 °C and stabilizes at 400 °C, with 24 wt.% EMBF$_4$

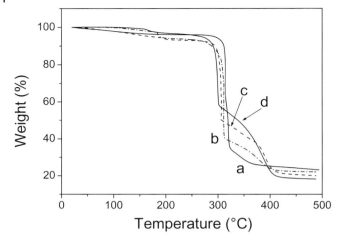

Figure 7.7 TGA graphs of PGE containing (a) 0, (b) 0.1, (c) 0.2, and (d) 0.3 mol of EMIBF$_4$.

remaining. In the case of the (b) curve, the composite containing 0.1 mol of EMIBF$_4$ showed an abrupt decreasing tendency at ~300 °C. From 320 °C to 400 °C, sample (b) showed a higher weight than sample (a), indicating that the thermal stability was improved by adding additional molten salts. Sample (c) showed an abrupt decay at ~300 °C to 50 wt.% EMBF$_4$, after which it decayed to 24 wt.% EMBF$_4$ until the temperature reached 400 °C. Finally, sample (d) showed a similarly abrupt decay at ~300 °C, to 50 wt.% EMBF$_4$, after which it slowly decayed to 24 wt.% EMBF$_4$ until the temperature reached 400 °C.

7.3.2.3 Electrochemical Properties

Impedance spectroscopy was used to determine the ionic conductivity of the PGE. The measurements were carried out at the 1 MHz to 1 Hz frequency range. The conductivity (σ) was measured by sandwiching the PGE films between the stainless steel electrodes and calculated from the bulk resistance (R_b), which was determined by equivalent circuit analysis software:

$$\frac{1}{\sigma} = \frac{R_b A}{t} \tag{7.2}$$

where t is the film thickness and A the film area.

The Li transference number, t_{Li}^+, was determined using a method combining dc polarization and ac impedance measurements, which has been reported in [55]. The equation being

$$t_{Li}^+ = \frac{I_S(\Delta V - I_O R_O)}{I_O(\Delta V - I_S R_S)} \tag{7.3}$$

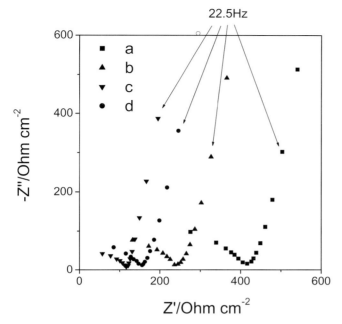

Figure 7.8 Impedance plots of PGE containing (a) 0, (b) 0.1, (c) 0.2, and (d) 0.3 mol of EMIBF$_4$.

where I is the dc current and R_o and R_s are the resistances of the passivation film formed onto the metallic lithium electrodes during the measurement. The subscripts o and s indicate the initial and steady states, respectively.

The current and resistance were measured using a potentiostat/galvanostat from AUTOLAB/PGSTAT30 (Eco Chemie, the Netherlands) and a Solatron 1260 Impedance Analyzer. The sample was sandwiched between two stainless steel disks with 0.5 mm-thick Li foils employed as nonblocking electrodes. A Li/PE film/Li cell was introduced into a coffee bag. The entire process was carried out in an argon-gas-filled glove box (M-Braun Lab Master 130), which was <1 ppm water and oxygen. Linear sweep voltammograms were taken at a scan of 10 mV/s, using the above-mentioned sandwiched samples.

In order to achieve commercial application feasibility, studies on the effects of the addition of specific amounts of molten salts on the ion-conducting behavior of the PGE systems were carried out.

Figure 7.8 shows the impedance plots versus molten salt contents for the PGE at 25 °C. An increase in the conductivity, as indicated by the decreased semicircle sizes in the higher frequency region, was achieved by adding molten salt. The maximum ion conductivity (i.e., the smallest semicircle on the impedance plot) was achieved at a content of 0.2 mol. When the EMIBF$_4$ was increased beyond 0.2 mol, the conductivity decreased slightly from the maximum value. It could be

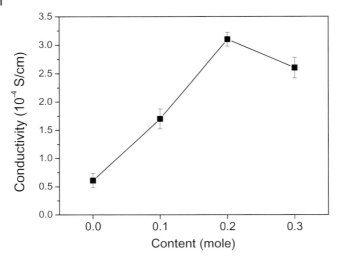

Figure 7.9 Room-temperature ion conductivity versus EMIBF$_4$ concentration (mol) for PGE.

concluded that the addition of the optimum EMIBF$_4$ content (0.2 mol) provided the most suitable environment for the ionic transportation and, thereby, achieved the highest conductivity. It was considered that the ion mobility of the PGE was enhanced due to the decreased crystallinity and beneficial features of the EMIBF$_4$ in comparison with those of the pristine electrolyte.

Figure 7.9 displays the ionic conductivity for the PGE as a function of the additional EMIBF$_4$ content. The conductivity increases with increasing EMIBF$_4$ content, and attains the maximum value, 3.1×10^{-4} S/cm, when the EMIBF$_4$ concentration is 0.2 mol. This is about five times the value, 6.1×10^{-5} S/cm, for the PEO/EC-LiBF$_4$ electrolytes. Subsequently, the ionic conductivity decreases slightly as the EMIBF$_4$ content increases beyond 0.2 mol. This decrease in the conductivity is mainly due to the decreased ion mobility; reflecting an increased PGE viscosity brought about by the high molten salt contents. The above DSC results imply, though, that the crystallinity is slightly decreased by addition of EMIBF$_4$ contents of up to 0.3 mol. However, the PGE with a content of 0.3 mol does not show the maximum conductivity. Hence, the change in ionic conductivity probably is not solely dependent on the crystalline structure change, but is also dependent on the ion mobility or interaction with the polymer chain or solvent species.

To examine the thermal dependence of the ionic conductivity of the PGE, AC impedance techniques were performed for the various PGEs at temperatures varying from 25 °C to 80 °C. The ionic conductivity changes as a function of EMIBF$_4$ content at those temperatures as shown in Figure 7.10. In the case of the PGE without molten salt, the ionic conductivity of the PGE increased, exhibiting a high-degree slope from 25 °C to 60 °C. After that point, the conductivity curve slope had been lowered until 80 °C. In the case of the PGE with a content of

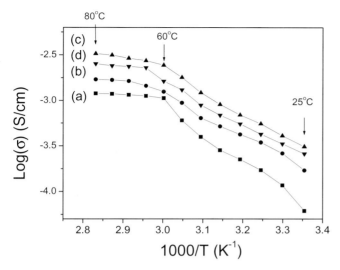

Figure 7.10 Ion conductivity versus $1000/T$ for PGE containing (a) 0, (b) 0.1, (c) 0.2, and (d) 0.3 mol of EMIBF$_4$.

0.1 mol, the conductivity increased gradually with a rather low-degree slope, compared with that of the above sample. The conductivity slope increase also changed abruptly at 60 °C. After that point, the conductivity behavior as a function of temperature was similar to that of the above sample. The PGE with a content of 0.2 mol EMIBF$_4$, which had the highest conductivity at room temperature, also showed a similar conductivity increase as a function of temperature and a slope change at 60 °C. Finally, the PGE with the content of 0.3 mol, which had a slight decrease in conductivity compared with the PGE with the content of 0.2 mol, showed a change of slope at a lower temperature of 65 °C.

For all of the PGE systems, the conductivity increased with increasing the temperature. A characteristic slope-change in the data was seen close to the polymer melting point, at almost 60 °C. Two regions, above and below the PEO melting point (T_m), were apparent. In both regions, the conductivity slowly increased with temperature, but a sudden slope change was observed across the melting point. According to previous reports, the change of conductivity with temperature was due to polymer segmental motion [56, 57]. The segmental motion either permits the ions to hop from one site to another, or provides necessary voids wherein the ions can move. Considering the above results, together with the DSC results, the ionic conductivity was influenced not only by the crystalline structure change but also by the ion mobility or interaction. In the case of the 0.2 mol EMIBF$_4$ content, it was possible that the interaction between the salts and the polymer matrix commenced, by which the ionic conductivity was enhanced. This could be attributed to the diminishment of the dilution effects of the lithium ions and the increment of the lithium ion transference rate.

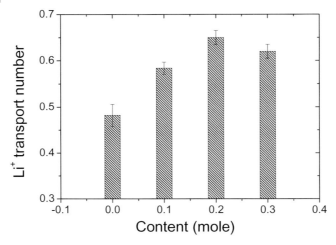

Figure 7.11 Lithium transference number of PGE as a function of the EMIBF$_4$ content.

The Li-ion transference number, t_{Li}^+, is one of the most important parameters for rechargeable Li-polymer batteries. A relatively high t_{Li}^+ can diminish the concentration gradients in the battery and thereby ensure battery operation under high-rate charge/discharge processes. Figure 7.11 shows the ion transference numbers for the various PGEs. In the case of the pristine PEO/EC/LiBF$_4$, the Li ions could interact not only with the ether oxygen in the PEO, but also with the BF$_4^-$, after which its transference capability was restricted, resulting in a low transference value (~0.48) [13]. For the PGE-containing EMIBF$_4$, the Lewis acid sites such as imidazolium cataion of the large-volume EMIBF$_4$ could interact with the oxygen elements in the PEO and BF$_4^-$ (Lewis base), and hence weaken the interactions between the PEO or salt and the Li ions. This tendency was similar to the one of additional general inorganic fillers, such as SiO$_2$ and Al$_2$O$_3$ to slightly increase the t_{Li}^+ through the well-known Lewis acid–base interactions [58, 59]. As a result, additional free Li ions are released, and t_{Li}^+ is enhanced. By increasing the EMIBF$_4$ content to 0.2 mol, the t_{Li}^+ is increased to ~0.65. This phenomenon might be explained by the increased degree of Lewis acid–base interactions between the polymer and EMIBF$_4$. The PGE with an EMIBF$_4$ content of 0.3 mol showed a slightly decreased t_{Li}^+ value, 0.58. This indicated that the above interaction was diminished by the heterogeneous morphology attendant on the decreased ion mobility.

Figure 7.12 shows the linear sweep voltammograms for the pristine PEO/EC/ LiBF$_4$ and EMIBF$_4$ salt-containing PGE. The irreversible onset of the current can be defined as the electrolyte breakdown voltage. The pristine sample shows a breakdown voltage of about 3.7 V versus Li. With the increase in the EMIBF$_4$ content, the PGE shows a linearly increased breakdown voltage. In the case of a content of 0.3 mol of EMIBF$_4$, the breakdown voltage extends to 4.5 V – a result of the interaction between EMIBF$_4$ and the PEO chains. It could be concluded that

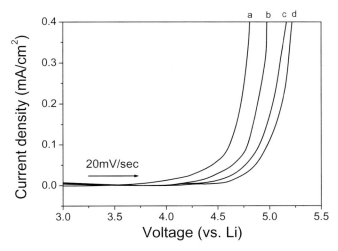

Figure 7.12 Linear sweep voltammograms of PGE as a function of the EMIBF$_4$ content.

the stable electrochemical window could be extended to 4.5 V versus Li by adding 0.3 mol of molten salt. The voltage limit was enhanced proportionally to the EMIBF$_4$ content. This tendency was somewhat different from the conductivity dependency on the molten salt content. In other words, the stable electrochemical voltage limit was solely dependent on the effect of the molten salt itself.

7.3.3
Characterization of Organic-Modified MMT-Containing Polymer Composite Electrolytes

Here three kinds of organic-modified MMT (OMMT) will be described, each of which varies from the rest according to the alkyl lengths and positions of the alkyl-ammonium functional group. Cloisite 20A (MMT-20A, MMT modified with a quaternary ammonium salt) was expected to have a large molecular volume size due to the two bulky groups of hydrogenated tallow (HT). Cloisite 25A (MMT-25A) has one bulky group of HT and a rather short-length alkyl group. Cloisite 30B (MMT-30B) has one bulky group of tallow and two short ethyl alcohol groups. The materials characterization was performed in the same manner as for the PEs of Section 7.3.2 above.

7.3.3.1 Morphologies and Structural Properties
Figure 7.13 shows the XRD patterns of (a) PEO/LiClO$_4$, and PCE containing (b) 10 wt.% Na-MMT, (c) 10 wt.% MMT-20A, (d) 10 wt.% MMT-25A, and (e) 10 wt.% MMT-30B in the $2\theta = 2$–$10°$ region. Pristine PEO/LiClO$_4$ had no characteristic peak in this region. Each PCE pattern had one peak, at $2\theta = 6.56$, 3.69, 3.82, and 4.07°, respectively. These peaks were assigned to the (001) lattice spacing of the

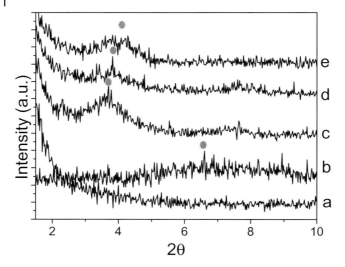

Figure 7.13 XRD patterns of (a) PEO/LiClO$_4$ and PCE containing 10 wt.% (b) Na-MMT, (c) MMT-20A, (d) MMT-25A, and (e) MMT-30B.

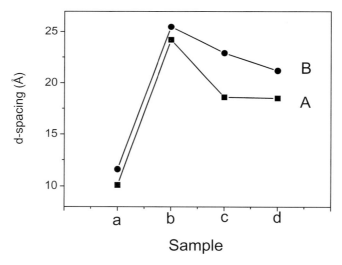

Figure 7.14 *d*-spacing lengths of (A) filler materials and (B) PCE containing fillers of (a) Na-MMT, (b) MMT-20A, (c) MMT-25A, and (d) MMT-30B.

MMT. As a result of the OMMT introduction, the *d*-spacing was largely enhanced compared with Na-MMT. However, each modified-filler-containing PCE showed a d_{001} peak, indicating that the layered MMT platelets had not been completely exfoliated (or delaminated) and had sustained the stacked structure with an expanded spacing.

Figure 7.14 illustrates the *d*-spacing length of the PCE for the different kinds of alkyl-ammonium modified MMT. Graph (A) shows the *d*-spacing lengths of the

Na-MMT, MMT-20A, MMT-25A, and MMT-30B, respectively. Among the samples, MMT-20A showed the highest *d*-spacing value owing to the large molecular volume of the dehydrogenated tallow functional group. Graph (B) shows the *d*-spacing lengths of the respective fillers after fabricating the PCE with PEO. The *d*-spacing lengths were increased, indicating that some parts of the polymer chain had been intercalated into the MMT interlayer. The PCE containing MMT-20A showed the highest *d*-spacing length, 25.5 Å, whereas the PCE containing Na-MMT showed the lowest, 11.6 Å. Furthermore, in the cases of (c) MMT-25A and (d) MMT-30B, the increased *d*-spacing length was rather large compared with that of the other samples, reflecting the fact that more of the polymer chain could be attracted and penetrated into the interlayer galleries.

Figure 7.15 shows the TEM results for the PCE containing (a) Na-MMT, (b) MMT-20A, (c) MMT-25A, and (d) MMT-30B. It is noticed that the qualitative OMMT *d*-spacing changes after preparation of the PCE. PCE/Na-MMT (a) showed the stacked platelet morphology of MMT. In the case of MMT-20A (b), the PCE showed rather expanded layered platelets compared with (a). It could be concluded that the MMT had not been fully exfoliated, which was confirmed by the existence

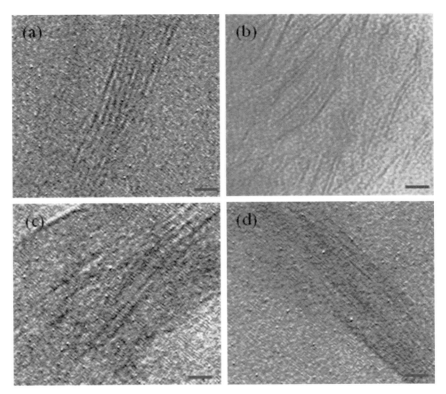

Figure 7.15 TEM images of PCE containing 10 wt.% fillers of (a) Na-MMT, (b) MMT-20A, (c) MMT-25A, and (d) MMT-30B (scale bar: 20 nm).

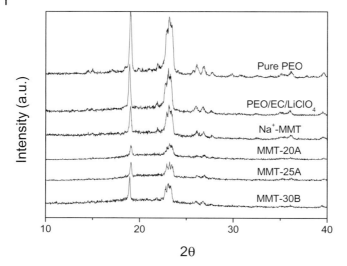

Figure 7.16 XRD patterns of (a) pure PEO, (b) PEO/EC/LiClO₄ and PCE containing 10 wt.% (c) Na⁺-MMT, (d) MMT-20A, (e) MMT-25A, and (f) MMT-30B.

of the d_{001} peak in the previous XRD data, as shown in Figure 7.13. PCE/MMM-25A (c) manifested the stacked platelet morphology of MMT, in which the interlayer spacing length was decreased in comparison with that of MMT-20A (b). PCE/MMT-30B (d) also showed the stacked morphology and the gap spacing was smaller than that of MMT-20A. In conclusion, these TEM studies, performed by the authors, confirmed that the d-spacing of MMT in PCE was enhanced and changed by using a different organic modifier. As shown in the TEM images, the MMT or OMMT was dispersed homogenously in the polymer matrix. Of course, some of the MMT existed as delaminated or exfoliated layers. However, most of the MMT layered-structures were well stabilized and somewhat expanded by organic modifiers or polymer chain intercalation. This expanded OMMT brought the modification of the microstructure of the PCE.

Figure 7.16 illustrates the wide-range XRD patterns of the various PEO/OMMT PCEs. These results showed the PCE crystallinity changes for different OMMT species. The characteristic diffraction peaks of the PEO crystalline structure were apparent between $2\theta = 15°$ and $30°$. These diffraction peaks became broader and less prominent with the various MMT additions, resulting in decreased crystallinity. By adding OMMT to the PEO/LiClO₄ system, the intensities of the characteristic peaks were decreased, the lowest value corresponding to MMT-20A.

7.3.3.2 Thermal Properties

The thermal parameters of the PCE, that is, the melting temperature and the heat of fusion (ΔH_f) are shown in Figure 7.17. PEO containing Na-MMT showed decreased melting temperature ($T_m = 52.4$ °C) and heat of fusion ($\Delta H_f = 82.4$ J/g);

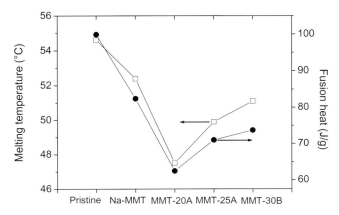

Figure 7.17 Thermal parameters of pristine PEO/LiClO₄ and PCE containing 10 wt.% Na⁺-MMT, MMT-20A, MMT-25A, and MMT-30B.

the corresponding values of pristine PEO are T_m = 54.6 °C, 99.9 J/g, meaning that the Na-MMT was effective in reducing the crystallinity of the PEO. By employing OMMT, the T_m and crystallinity of the PCE were effectively reduced further than those of the PCE containing Na-MMT. The lowest melting temperature (47.5 °C) and heat of fusion (62.5 J/g) were obtained for the PCE containing MMT-20A. Considered with the previous XRD results, the DSC studies indicated that the PEO crystalline structure had deteriorated and that the reduced crystallinity was dependent on the d-spacing of the OMMT. As a result, the interlayer spacing of the MMT was enhanced by using organic modifiers that had a large molecular volume. It was thought that the expanded MMT layer silicates might interact effectively with a polymer chain. By increasing the filler–matrix interaction, the PCE film morphology or crystalline structure changed. The filler effect was enhanced by the expanded MMT by controlling the organic modifiers.

7.3.3.3 Electrochemical Properties

Figure 7.18 shows the conductivity for the PCE containing various types of MMT at 25 °C. The highest ion conductivity, 6.1×10^{-4} S/cm, was observed for MMT-20A. This was about triple the value, 2.2×10^{-4} S/cm, of PEO/Na-MMT, indicating that the ion conductivity of the PCE was enhanced by increasing the interlayer spacing of the MMT clay.

Figure 7.19 shows the impedance plots versus MMT contents for the PEO/MMT-20A PCE at 25 °C. An increase in the conductivity, as indicated by the decreased semicircle in the higher frequency region, was achieved by adding a small quantity of the OMMT. The maximum ion conductivity (i.e., the smallest semicircle on the impedance plot) was achieved at 10 wt.% OMMT content. When the OMMT was increased beyond 10 wt.%, the conductivity decreased slightly from the maximum value. It could be concluded that the addition of the optimum OMMT content (10 wt.%) provided the most suitable environment for the ionic

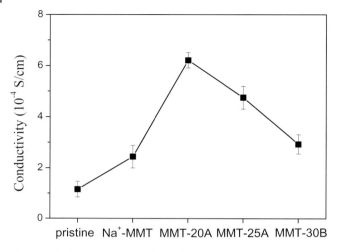

Figure 7.18 Ion conductivity values of PEO/LiClO₄ and PCE containing 10 wt.% fillers of (a) Na-MMT, (b) MMT-20A, (c) MMT-25A, and (d) MMT-30B.

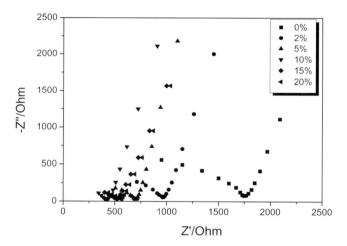

Figure 7.19 Electrochemical impedance spectra of PCE containing MMT-20A with several different weight contents (wt.%).

transportation and, thereby, achieved the highest conductivity. By increasing the interlayer spacing of the MMT, the crystalline structure of the PCE degraded, meaning that the crystallinity decreased, leading to a more flexible chain structure of the PC. Consequently, lithium ions could move more easily along the flexible polymer main chain, resulting in the enhancement of ion conductivity. This increase in ion conductivity by OMMT addition could be related to the functioning of a few MMT clay sheets that were expanded and dispersed in the matrix.

Such functions have been explained, by the reference to cation interactions, in a previous report [60]. These authors suggested three types of Li cation complexes, those with (a) polyether chains, (b) polyether and silicate layers, and (c) silicate layers. For all types of interaction, the interfacial area among the Li ions, polyether chains, and silicate layers would be critical to the filler effects. In this view, the larger *d*-spacing of the OMMT could probably afford the much higher interaction of fillers with polymer chains or Li ions.

7.3.4
Ion-Exchanged Li-MMT-Containing Polymer Composite Electrolytes

7.3.4.1   Structural Properties
Figure 7.20 shows the diffraction patterns of the small-angle region for PCE as a function of Li-MMT contents. The interlayer spacing value was changed from 32.43 to 25.66 Å when the content was increased from 10 to 25 wt.%. The spacing of the samples progressively decreased with Li-MMT until a maximum was reached, which in fact was related to the difficulty of intercalating PEO molecules into Li-MMT due to the increased content of clay fillers. In other words, the concentration of PEO-intercalated Li-MMT decreased with the increase in Li-MMT content. It is worth pointing out that PEO/MMT blends are nanosized composites, and it is expected that these composites exhibit silica layers with a marked specific surface area.

Figure 7.21 displays the crystallinity changes in the PCE with Li-MMT by XRD. The characteristic diffraction peaks of the crystalline PEO are apparent between

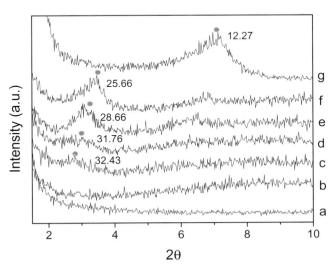

Figure 7.20 X-ray diffraction patterns of (a) pure PEO, PCE containing (b) 5%, (c) 10%, (d) 15%, (e) 20%, (f) 25% of Li-MMT, and (g) Li-MMT itself.

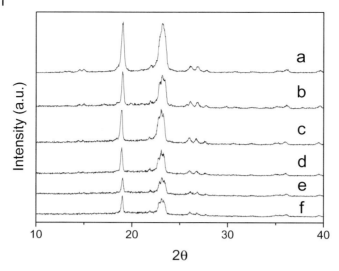

Figure 7.21 X-ray diffraction patterns of (a) pure PEO and PCE containing (b) 0%, (c) 5%, (d) 10%, (e) 20%, and (f) 25% of Li-MMT.

$2\theta = 15°$ and $30°$ [52, 61]. These diffraction peaks become broader and less prominent with increasing Li-MMT content, resulting in decreased crystallinity. With the addition of the Li-MMT to the PEO/EC system, the intensities of characteristic peaks decreased, and reached the lowest value at 20 wt.%.

7.3.4.2 Thermal Properties
Figure 7.22 shows the DSC results of PEO/EC electrolytes with different contents of Li-MMT. The endothermic peaks at the melting temperature, attributed to the melting of the PEO-rich crystalline phase, weakened and shifted to lower temperatures with increasing the Li-MMT content up to 20 wt.%.

Both the XRD and DSC studies indicate that PEO crystallinity is deteriorated by the EC and further by the addition of Li-MMT. The melting temperature, heat of fusion, and crystallinity are exhibited in Table 7.3. The PEO itself shows the heat of fusion as 189.93 J/g, and PEO/EC (2 : 1) without filler shows the value as 102.50 J/g, the estimated crystallinity being 53.97%. This indicates that EC can reduce the crystallinity of the PEO by plasticizer effects. With the increase in filler content to 20 wt.%, the melting point gradually fell to 52.2 °C, and the crystallinity value fell to 37.73%. In the case of 25 wt.%, the PCE showed the slightly increased melting point of 54.80 °C and the enhanced crystallinity of 38.74%.

TGA graphs of the Li-MMT and PCE are shown in Figure 7.23. PEO itself generally begins to decompose at 200–300 °C, and is completely burned away above 350 °C under a N_2 atmosphere. The weight loss of the Li-MMT itself until 200 °C can be attributed to water evaporation or removal. A dried Li-MMT particle contains about 12% moisture, most of which is lost below 100 °C. Above 100 °C, the

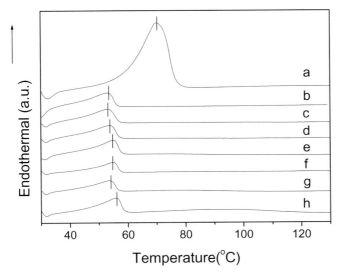

Figure 7.22 DSC curves of (a) pure PEO and PCE containing (b) 0%, (c) 2%, (d) 5%, (e) 10%, (f) 15%, (g) 20%, and (h) 25% of Li-MMT.

Table 7.3 DSC data for PCE containing various Li-MMT contents.

Li-MMT content (wt.%)	Melting temp., T_m (°C)	Heat of fusion, ΔH_f (J/g)	Crystallinity, χ (%)
Pure PEO	65.4	189.93	–
0	53.2	102.50	53.97
2	52.9	98.56	51.89
5	52.7	94.67	49.67
10	54.2	91.33	48.09
15	52.5	90.67	47.74
20	52.2	71.67	37.73
25	54.8	73.59	38.74

water molecules adhesively adhere to the interlayer of the silicates and begin to evaporate. It is clear that the PEO/Li-MMT composites contain less moisture than this. This explains the fact that the water molecules located between silicate sheets are easily exchanged for polymer chains in the composites. Above 400 °C, the remaining weight of the PCE increased linearly as the Li-MMT content increased indicating that the thermal stability was improved by adding Li-MMT.

7.3.4.3 Electrochemical Properties
Figure 7.24 shows the ionic conductivity for Li-MMT in 2:1 mixtures of PEO and EC. The conductivity increases with an increase in Li-MMT content and attains a maximum value of 5.3×10^{-6} S/cm when the Li-MMT content is 20 wt.%.

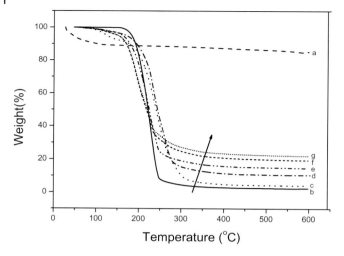

Figure 7.23 TGA graphs for (a) Li-MMT, (b) PEO itself, and PCE containing (c) 0%, (d) 10%, (e) 15%, (f) 20%, and (g) 25% of Li-MMT.

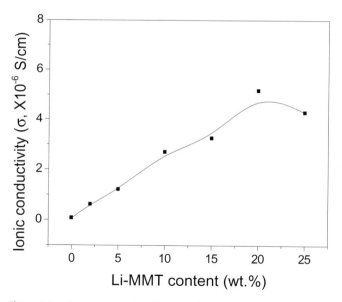

Figure 7.24 Room temperature ionic conductivity of PCE with Li-MMT of different weight ratios.

Subsequently, the ionic conductivity decreases slightly with further increases in Li-MMT content over 20 wt.%. These decreases in the conductivity are mainly due to the increased crystallinity reflecting an aggregation of additives at high filler contents [46]. On the other hand, the above DSC results also imply that ionic

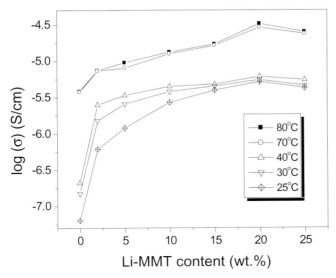

Figure 7.25 Ionic conductivity versus Li-MMT concentration (wt.%) for PCE at different temperatures.

conductivity is strongly dependent on the degree of crystallinity. Ionic conduction takes place mainly through amorphous domains. Hence, the level of ionic conductivity below the melting temperature reflects the degree of crystallinity of a sample. Higher crystallinity is known to yield a lower conductivity if the other conditions are the same [62].

The ionic conductivity changes as a function of Li-MMT content at various temperatures ranging from 25 °C to 80 °C, as shown in Figure 7.25. At the ambient temperature of 25 °C, the ionic conductivity of the PCE increased, exhibiting a high-degree slope from 0 to 5 wt.% Li-MMT content. After that point, the conductivity increased gradually, exhibiting a rather low-degree slope at 20 wt.% content. In the case of 25 wt.%, the conductivity decreased slightly, compared with that of 20 wt.% content. The tendency of conductivity to abruptly increase before reaching 5 wt.% content was also observed at 30 °C and 40 °C. However, this tendency disappeared in the cases of 70 °C or 80 °C temperatures. The conductivity behavior as a function of filler content differed between 40 °C and 70 °C. Furthermore, the conductivity curves as a function of filler content increased in degree, to one order of magnitude, from 40 °C to 70 °C. This is probably related to the phase transition (i.e., melting) of PEO.

Figure 7.26 shows the Arrehenius plots for the PCE as a function of Li-MMT content. For all PCE systems, conductivity increases with increasing temperature. There exists a large change of slope at 3.00 (1000/T) (60 °C). A characteristic slope-change in the data for the samples is seen close to the polymer melting point, almost 60 °C. Two regions, above and below the PEO melting point, are apparent. In both regions, the conductivity slowly increases with temperature, but is observed

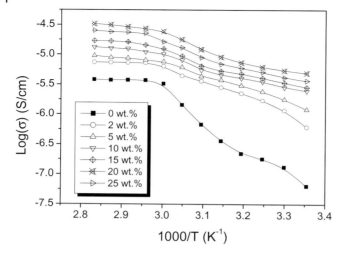

Figure 7.26 Arrehenius plots for the PCE as a function of Li-MMT content.

Table 7.4 Activation energies for PCE with different weight ratios of Li-MMT.

Li-MMT content (wt.%)	Above 60 °C (eV)	Below 60 °C (eV)
0	0.033	0.70
2	0.028	0.18
5	0.047	0.16
10	0.052	0.15
15	0.061	0.16
20	0.056	0.19
25	0.089	0.16

as a sudden jump across the melting point. According to previous reports, the change of conductivity with temperature is due to polymer segmental motion [56, 57]. The segmental motion either permits the ions to hop from one site to another or provides necessary voids for ions to move.

The linear variation in the (σ vs. $1/T$) plots below and above T_m suggests an Arrhenius-type thermally activated process. In both regions, the conductivity relationships can be expressed as

$$\sigma = \sigma_0 \exp(-E_a/kT) \tag{7.4}$$

where σ_o is the pre-exponential factor, E_a is the activation energy, and k is the Boltzman constant.

The activation energies for ion conductivity calculated from the Arrhenius relationships are summarized in Table 7.4. The activation energy values above 60 °C for all of the samples are shown to be reduced when compared with those below

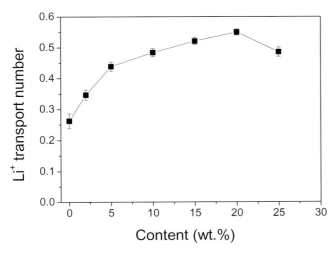

Figure 7.27 Li-ion transference number as a function of Li-MMT content.

60 °C. Below 60 °C, the values decreased as a result of the addition of Li-MMT, and showed the lowest when the Li-MMT content was 10%. However, above 60 °C, the effect of the addition of Li-MMT was rather small.

Li-ion transference numbers, t_{Li}^+, as a function of Li-MMT content are shown in Figure 7.27. For the PCE containing Li-MMT, the Lewis acid sites on the anionic surface of Li-MMT can interact with oxygen elements in the PEO and ClO_4^- (Lewis base), and hence weaken the interactions between the oxygen elements and the Li ions. As a result, more free Li ions are released, and the t_{Li}^+ is enhanced. This is an identical tendency to that by which the addition of general inorganic fillers, such as SiO_2 and Al_2O_3, slightly increases the t_{Li}^+ through the well-known Lewis acid–base interactions [58, 59]. By increasing the Li-MMT content, the t_{Li}^+ is increased to a value, 0.55, until 20 wt.% Li-MMT content is reached, and then it is slightly decreased. Over 20% content, the above interfacial area would be diminished by a heterogeneous morphology due to the filler aggregation.

7.3.5
Mesoporous Silicate (MCM-41)-Containing Polymer Composite Electrolytes

7.3.5.1 Morphologies and Structural Properties
The hexagonal pore structure and pore diameter of the MCM-41 confirmed by TEM is shown in Figure 7.28. The TEM image verified the average pore diameter of 25.8 Å obtained by the isotherm experiment.

Also, the morphologies of PEO/MCM-41 were demonstrated using the SEM technique, as shown in Figure 7.29. The pristine PEO-LiClO₄ shows some white heterogeneous spots, which denote the crystalline regions. These white spots slightly disappeared with the increase of MCM-41 content, indicating that the

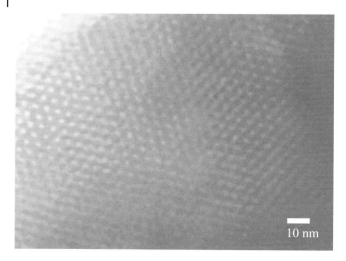

Figure 7.28 TEM image indicating the hexagonal pore structure and pore diameter of the MCM-41.

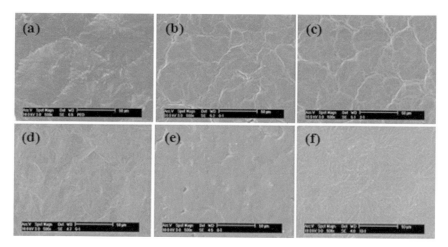

Figure 7.29 SEM images of PCE containing (a) 0, (b) 2, (c) 5, (d) 8, (e) 10, and (f) 15 wt.% MCM-41.

PCE morphology became homogenous due to the decrease of crystallinity. This result confirms that the PCE was successfully prepared as a homogenous nanocomposite.

Figures 7.30(A) and (B) show the small-angle XRD patterns of the as-prepared MCM-41 and PCE containing the MCM-41, respectively. The four peaks were attributed to the hexagonal structure of the MCM-41 pores. The (100) peak at $2\theta = 2.3°$ showed the strongest intensity, the others being assigned to [110], [200],

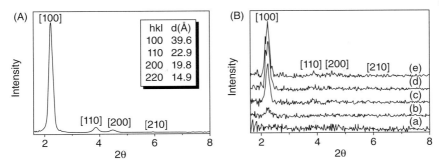

Figure 7.30 Small-angle X-ray diffraction patterns of (A) MCM-41 itself, and (B) PEO–LiClO₄ PCE containing (a) 2%, (b) 5%, (c) 8%, (d) 10%, and (e) 15% MCM-41.

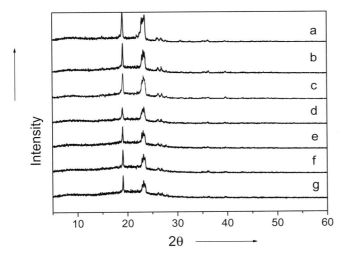

Figure 7.31 Wide-angle X-ray diffraction patterns of (a) pure PEO and PCE containing (b) 0%, (c) 2%, (d) 5%, (e) 8%, (f) 10%, and (g) 15% MCM-41.

and [210] [63]. In Figure 7.30(B), with the increase in MCM-41 content, the peak at $2\theta = 2.3°$ showed a gradually increasing intensity. In the case of 5 wt.% content, the small characteristic peaks of the MCM-41 itself appeared. The crystalline structure of the MCM-41 could be confirmed by an intense peak at $2\theta = 2.3°$ and three small peaks at $3° < 2\theta < 6°$. These peaks could be assigned to [100], [110], [200], and [220], respectively. The peak positions did not change, indicating that most of the MCM-41 could maintain their ordered-pore structures effectively and be homogeneously complexed with the PEO chains. It is worth pointing out that PEO/MCM-41 are nanosized composites having ~2.6 nm pore size, and it is expected that these composites exhibit silicate layers of a specific surface area.

Figure 7.31 displays the wide-angle XRD pattern changes in the PCE containing MCM-41. The characteristic diffraction peaks of the PEO crystalline phase are

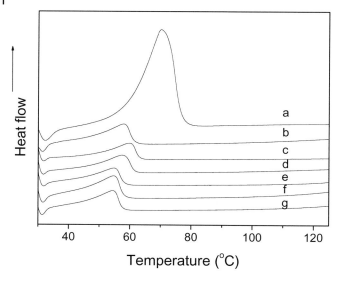

Figure 7.32 DSC curves of (a) pure PEO, and PCE containing (b) 0%, (c) 2%, (d) 5%, (e) 8%, (f) 10%, and (g) 15% MCM-41.

apparent between $2\theta = 15°$ and $30°$ [52, 61]. These diffraction peaks become broader and less prominent with increasing MCM-41 concentration, resulting in decreased crystallinity.

7.3.5.2 Thermal Properties

To confirm the effect of MCM-41 addition on the crystalline structure of the PCE, a crystallinity value was obtained by the DSC method, as shown in Figure 7.32. With the addition of the MCM-41 to the PEO–LiClO$_4$ system, the intensities of the characteristic peaks decreased, reaching the lowest value at 8 wt.%. The melting transition, corresponding to the crystalline-to-amorphous transition, was very sluggish, and the endothermic reaction was shown to start below the melting point [62].

MCM-41 fillers can decrease the crystallinity of PEO by the following means. One is through the Lewis acid–base interactions between the ether oxygen of PEO and the Lewis acid sites on the surface of MCM-41, as in the cases of SiO$_2$ and Al$_2$O$_3$. The other is by preventing the reorganization of PEO chains that partly intercalate into the mesoporous pore channels during the fabrication of the PCE.

The value χ has been defined as the enthalpy ratio of PEO samples to the complete crystalline PEO. It can be calculated from the equation

$$\chi = \Delta H_f / \Delta H_f^\circ \times 100 \tag{7.5}$$

where, ΔH_f° (213.7 J/g) is the melting enthalpy of a completely crystalline PEO sample and ΔH_j is the experimental enthalpy [53].

Table 7.5 DSC data for PCE containing various MCM-41 contents.

MCM-41 content (wt.%)	Melting temp., T_m (°C)	Heat of fusion, ΔH_f (J/g)	Crystallinity, χ (%)
Pure PEO	65.4	189.9	88.9
0	56.9	96.8	45.3
2	58.8	87.8	41.1
5	56.3	81.0	37.9
8	53.4	65.0	30.4
10	53.1	68.6	32.1
15	52.6	71.8	33.6

The crystallinity value can describe the relative change of the crystalline or amorphous phase of PCEs. The melting temperature and melting enthalpy, and the resulting crystallinity values, are summarized in Table 7.5. The melting temperature and crystallinity decreased when the content of MCM-41 increased. PEO-LiClO$_4$ itself showed a melting point of 56.9 °C and a heat of fusion of 96.8 J/g, implying the crystallinity of 45.3%. With the increase in MCM-41 content to 8 wt.%, the melting temperature gradually fell to 53.4 °C, and the crystallinity fell to the lowest value, 30.4%. In the case of 15 wt.%, the PCE showed a melting point of 52.6 °C and an enhanced crystallinity of 33.6%. With over 8 wt.% MCM-41, the crystallinity slightly increased, suggesting that there was an optimum content for the minimization of crystallinity. This increase in crystallinity might be related to the aggregation of MCM-41, resulting in the increased crystalline domain in the composite matrix. The decrease in crystallinity will increase the amorphous state of PEO, which should enhance the ion conductivity in low-temperature regions.

7.3.5.3 Electrochemical Properties

Complex impedance plots for PCEs as a function of MCM-41 content are shown in Figure 7.33. In the case of 2 wt.% MCM-41, the graph showed a rather large-diameter semicircle. The approximate bulk-resistance value could be obtained with reference to the touch point at the *x*-axis. By contrast, the PCE containing 8 wt.% MCM-41 showed the smallest diameter semicircles.

Figure 7.34 displays the effect of the MCM-41 content on the enhancement of the room-temperature ion conductivity of the PCE. The enhancement of ion conductivity first increases sharply with MCM-41 content, and reaches the maximum conductivity of 1.2×10^{-4} S/cm at 8 wt.% content. This value is two-orders-of-magnitude higher than that of the pristine PEO–LiClO$_4$. This conductivity value is at a slightly higher or similar level with the previous results [64–66]. But, when the MCM-41 content increased further, the ion conductivity decreased. Thus, when the content of MCM-41 is high, over 8 wt.%, the blocking effect on the transport of charge carriers, perhaps resulting from the aggregation of the MCM-41, can decrease the ion conductivity of the PCE. Besides, the decrease in the conductivity for MCM-41 contents higher than 8 wt.% can be caused not only by

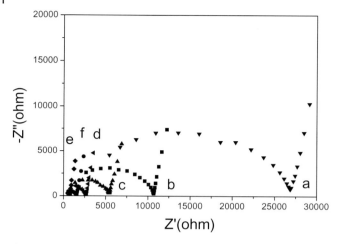

Figure 7.33 Complex impedance plots of PCE with MCM-41 of different weight ratios.

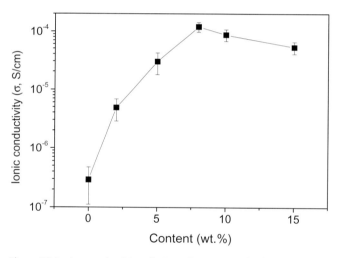

Figure 7.34 Ion conductivity of PCE with MCM-41 of different weight ratios at room temperature.

aggregation but also by the dilution effect of introducing a nonconducting phase in a conducting one. Alternatively, the above DSC results also imply, once again, that ionic conductivity is dependent on the varying degree of crystallinity. Ionic conduction takes place mainly through the amorphous domains; hence, the level of ionic conductivity below the melting point reflects the degree of crystallinity of the PCE [47, 48].

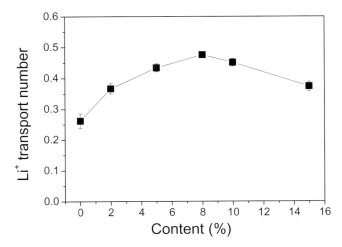

Figure 7.35 Li-ion transference number for PCE as a function of MCM-41 content.

Li-ion transference numbers, t_{Li}^+, as a function of MCM-41 content are shown in Figure 7.35. As previously mentioned a relatively high t_{Li}^+ can ensure proper battery operation under a high-rate charge/discharge process.

In the case of pristine PEO–LiClO$_4$, Li ions can interact not only with ether oxygen in PEO but also with the oxygen element in ClO$_4^-$ (Lewis base), in which case its transport ability is restricted, resulting in a low value ($<$0.3) [14]. This is an identical tendency to that by which the addition of general inorganic fillers, such as SiO$_2$ and Al$_2$O$_3$, slightly increases the t_{Li}^+ through the well-known Lewis acid–base interactions [57, 59]. For PEO-LiClO$_4$ containing MCM-41, the Lewis acid sites on the surface of MCM-41 can interact with oxygen elements in PEO and ClO$_4^-$, and hence weaken the interactions between the oxygen elements and then Li ions. As a result, more free Li ions are released, and the t_{Li}^+ is enhanced. By increasing the MCM-41 content, the t_{Li}^+ is increased to a value, 0.5, until 8 wt.% MCM-41 is reached, and then it is slightly decreased. This phenomenon could be explained by an increased degree of Lewis acid–base interactions resulting from the enhanced interfacial area between the polymer and MCM-41. In contrast, the above interfacial area would be diminished by heterogeneous morphology due to the filler aggregation, as already mentioned in Li–MMT fillers.

7.4
Conclusions

The present study demonstrated that additives for PEs such as RTMS, inorganic fillers, organic plasticizers, play a major role to decrease the crystallinity of PEO-based CEs and thereby function as a Li-ion transference improvement agent. In order to measure the enhanced ion-conducting behaviors of PEs, the effects of the additive content on the structural and electrochemical properties of the PE were investigated and the following conclusions can be drawn:

1) The crystallinity of the PEO can be decreased mainly by the structural deterioration that takes place due to the addition of EC and Li salt; crystallinity is further reduced slightly by adding EMIBF$_4$, OMMT, Li–MMT, and MCM-41.

2) The ionic conductivity of the PE gradually increased with increasing the additives. (a) The increase in ionic conductivity of molten-salt-containing PGEs as a function of EMIBF$_4$ concentration could be related to the presence of large-volume salt species dispersed in the matrix, resulting in a 27.9% decreased crystallinity. (b) The increase in ionic conductivity of inorganic fillers containing PCEs as a function of filler concentration could be related to the presence of the fillers being well dispersed in the matrix, resulting in decreasing the crystallinity of the PEO.

3) According to the results for the Li transference number, t_{Li}^+ increased initially as the additive content increased, but with a continuous increase in the additive content, t_{Li}^+ slightly decreased. This phenomenon could be explained by the occurrence of an enhanced degree of Lewis acid–base interaction between the polymer and the additives. For a more detailed explanation, Lewis acid sites on the surface of additives could interact with oxygen elements in PEO and ClO$_4^-$, and hence weaken the interactions between the oxygen elements and the Li ions. As a result, additional free Li ions were released and the t_{Li}^+ was enhanced.

4) This kind of layered-structured fillers, due to the regularly ordered structures, also could enhance the mechanical properties of the PE; good tensile and high modulus properties are preferred for the film-making process. These mechanical properties are currently under investigation.

PEs are comprised of metal salts of low lattice energy dissolved in a polymer matrix containing polar moieties such as ether, ester, or amide linkages. Metal salts can dissolve into such polymer matrixes by virtue of the coordinative interaction between the metal ion and the polar groups. The conductive properties of PEs have resulted in their extensive study for potential applications in energy storage devices or electrochromic devices, especially for the dye-sensitized solar cells. In the latter case, the ionic conductivity of solid PEs is dependent on the molar ratio of the polymer and the iodide salt due to the transfer efficiency of charge carriers and complex formation between metal cations and polymer atoms. To summarize this chapter, it can be said that PEs stand out because of their excellent properties such as easy fabrication, low cost, and good stability so that they have growing interests from the point of practical applications. It is expected that the additive effect of PEs for lithium secondary batteries could be useful for the study of enhancing the ionic conductivity of PEs for other applications as well.

References

1 Chawla, K.K. (1987) *Composite Materials: Science and Engineering*, Springer, New York.
2 Vincent, C.A. and Scrosati, B. (eds) (1997) *Modern Batteries: An Introduction to Electrochemical Power Sources*, Butterworth-Heinemann, London.
3 Britz, J., Meyer, W.H., and Wegner, G. (2007) Blends of poly(meth)acrylates with 2-oxo-(1,3)dioxolane side chains and lithium salts as lithium ion conductors. *Macromolecules*, **40** (21), 7558–7565.
4 Brandt, K. (1994) Historical development of secondary lithium batteries. *Solid State Ion.*, **69** (3–4), 173–183.
5 Cameron, G.G., Ingram, M.D., and Sarmouk, K. (1990) Conductivity and viscosity of liquid polymer electrolytes plasticized by propylene carbonate and tetrahydrofuran. *Eur. Polym. J.*, **26** (10), 1097–1101.
6 Scrosati, B. (1994) *Electrochemistry of Novel Materials* (eds J. Lipkowski and P.N. Ross), Wiley-VCH, Weinheim, Chapter 3.
7 Koksberg, R., Barker, J., Shi, H., and Saidi, M.Y. (1996) Cathode materials for lithium rocking chair batteries. *Solid State Ion.*, **84** (1–2), 1–21.
8 Gray, F.M. (1991) *Solid Polymer Electrolytes* (ed. F.M. Gray), Wiley-VCH, New York, Chapter 2.
9 Dias, F.B., Plomp, L., and Veldhuis, J.B.J. (2000) Trends in polymer electrolytes for secondary lithium batteries. *J. Power Sources*, **88** (2), 169–191.
10 Kim, S., Jung, Y., and Park, S.J. (2007) Effect of imidazolium cation on cycle life characteristics of secondary lithium–sulfur cells using liquid electrolytes. *Electrochim. Acta*, **52** (5), 2116–2122.
11 Kim, S., Jung, Y., and Park, S.J. (2005) Effects of imidazolium salts on discharge performance of rechargeable lithium–sulfur cells containing organic solvent electrolytes. *J. Power Sources*, **152**, 272–277.
12 Callum, J.R.M. (1987) *Polymer Electrolyte Reviews*, Elsevier, Amsterdam.
13 Baril, D., Michot, M., and Armand, M. (1997) Electrochemistry of liquids vs. solids: polymer electrolytes. *Solid State Ion.*, **94** (1–4), 35–47.
14 Croce, F., Appetecchi, G.B., Persi, L., and Scrosati, B. (1998) Nanocomposite polymer electrolytes for lithium batteries. *Nature*, **394** (6692), 456–458.
15 Kim, S. and Park, S.J. (2009) Preparation and electrochemical properties of composite polymer electrolytes containing 1-ethyl-3-methylimidazolium tetrafluoroborate salts. *Electrochim. Acta*, **54** (14), 3775–3780.
16 Liu, M., Visco, S.J., and De Jonghe, L.C. (1991) Novel solid redox polymerization electrodes. *J. Electrochem. Soc.*, **138** (7), 1891–1895.
17 Chu, M.Y., De Jonghe, L.C., and Visco, S. (eds) (1996) High specific power lithium polymer rechargeable battery. Proceedings of the Annual Battery Conference on Applications and Advances, p. 163.
18 Oman, H. (1997) High specific power lithium polymer rechargeable battery. Proceedings of the Annual Battery Conference on Applications and Advances, p. 31.
19 Ulbralife Batteries Inc. (1996) Ultralife Batteries, Product Data Sheet, published on the internet. www.ulbi.com (accessed June 2009).
20 Turrentine, T. and Kurani, K. (1995) Inst. of Transportation Studies. Report UCD-ITS-RR-95-5.
21 Mader, J. (1996) Commercialization of Advanced Batteries, IEE report, No. 0-7803-2994-5/96.
22 Fenton, D.E., Parker, J.M., and Wright, P.V. (1973) Complexes of alkali metal ions with poly(ethylene oxide). *Polymer*, **14** (11), 589–589.
23 Meyer, W.H. (1998) Polymer electrolytes for lithium-ion batteries. *Adv. Mater.*, **10** (6), 439–448.
24 Tarascon, J.M. and Armand, M.B. (2001) Issues and challenges facing

rechargeable lithium batteries. *Nature*, **414** (6861), 359–367.

25 Kim, Y.W., Lee, W., and Choi, B.K. (2000) Relation between glass transition and melting of PEO–salt complexes. *Electrochim. Acta*, **45** (8–9), 1473–1477.

26 Zhou, H., Gu, N., and Dong, S. (1998) Studies of ferrocene derivative diffusion and heterogeneous kinetics in polymer electrolyte by using microelectrode voltammetry. *J. Electroanal. Chem.*, **441** (1–2), 153–160.

27 Kuo, H.H., Chen, W.C., Wen, T.C., and Gopalan, A. (2002) A novel composite gel polymer electrolyte for rechargeable lithium batteries. *J. Power Source*, **110** (1), 27–33.

28 Fuller, J., Carlin, R.T., and Osteryoung, R.A. (1997) The room temperature ionic liquid 1-ethyl-3-methylimidazolium tetrafluoroborate: electrochemical couples and physical properties. *J. Electrochem. Soc.*, **144** (11), 3881–3885.

29 Yoshizawa, M., Ogihara, W., and Ohno, H. (2001) Design of new ionic liquids by neutralization of imidazole derivatives with imide-type acids. *Electrochem. Solid-State Lett.*, **4** (6), E25–E27.

30 Koch, V.R., Nanjundiah, C., Appetecchi, G.B., and Scrosati, B. (1995) The interfacial stability of Li with 2 new solvent-free ionic liquids – 1,2-dimethyl-3-propylimidazolium imide and methide. *J. Electrochem. Soc.*, **142** (7), L116–L118.0.

31 McEwen, A.B., McDevitt, S.F., and Koch, V.R. (1997) Nonaqueous electrolytes for electrochemical capacitors: imidazolium cations and inorganic fluorides with organic carbonates. *J. Electrochem. Soc.*, **144** (4), L84–L86.

32 Balducci, A., Bardi, U., Caporali, S., Mastragostino, M., and Soavi, F. (2004) Ionic liquids for hybrid supercapacitors. *Electrochem. Commun.*, **6** (6), 566–570.

33 Sakaebe, H. and Matsumoto, H. (2003) *N*-Methyl-*N*-propylpiperidinium bis(trifluoromethanesulfonyl)imide (PP13–TFSI) – novel electrolyte base for Li battery. *Electrochem. Commun.*, **5** (7), 594–598.

34 Sato, T., Maruo, T., Marukane, S., and Takagi, K. (2004) Ionic liquids containing carbonate solvent as electrolytes for lithium ion cells. *J. Power Source*, **138** (1–2), 253–261.

35 Plichta, E.J. and Behl, W.K. (2000) A low-temperature electrolyte for lithium and lithium-ion batteries. *J. Power Sources*, **88** (2), 192–196.

36 Berhier, C., Gorecki, W., Minier, M., Armand, M.B., Chabagno, J.M., and Riquaud, P. (1983) Microscopic investigation of ionic conductivity in alkali metal salts-poly(ethylene oxide) adducts. *Solid State Ion.*, **11** (1), 91–95.

37 Leo, C.J., Lao, G.V.S., and Chowdari, B.V.R. (2002) Studies on plasticized PEO–lithium triflate–ceramic filler composite electrolyte system. *Solid-State Ion.*, **148** (1–2), 159–171.

38 Appetecchi, G.B., Scaccia, S., and Passerini, S. (2000) Investigation on the stability of the lithium-polymer electrolyte interface. *J. Electrochem. Soc.*, **147** (12), 4448–4452.

39 Rhoo, H.J., Kim, H.T., Park, J.K., and Hwang, T.S. (1997) Ionic conduction in plasticized *PVC/PMMA* blend polymer electrolytes. *Electrochim. Acta*, **42** (10), 1571–1579.

40 Starkey, S.R. and Frech, R. (1997) Plasticizer interactions with polymer and salt in propylene carbonate-poly(acrylonitrile)-lithium triflate. *Electrochim. Acta*, **42** (3), 471–474.

41 Nookala, M., Kumar, B. and Rodrigues, S. (2002) Ionic conductivity and ambient temperature Li electrode reaction in composite polymer electrolytes containing nanosize alumina. *J. Power Sources*, **111** (1), 165–172.

42 Ennari, J., Pietila, L., Virkkunen, V., and Sundholm, F. (2002) Molecular dynamics simulation of the structure of an ion-conducting PEO-based solid polymer electrolyte. *Polymer*, **43** (20), 5427–5438.

43 Kim, S., Hwanga, E.J., Jung, Y., Han, M., and Park, S.J. (2008) Ionic conductivity of polymeric nanocomposite electrolytes based on poly(ethylene oxide) and organo-clay materials. *Colloids Surf. A*, **313-314**, 216–219.

44 Fu, X.A. and Qutubuddin, S. (2005) Swelling behavior of organoclays in styrene and exfoliation in nanocompos-

ites. *J. Colloid Interface Sci.*, **283** (2), 373–379.

45 Qian, X., Gu, N., Cheng, Z., Yang, X., Wang, E., and Dong, S. (2001) Impedance study of (PEO)$_{10}$LiClO$_4$–Al$_2$O$_3$ composite polymer electrolyte with blocking electrodes. *Electrochim. Acta*, **46** (12), 1829–1836.

46 Kim, S. and Park, S.J. (2007) Preparation and ion-conducting behaviors of poly(ethylene oxide)-composite electrolytes containing lithium montmorillonite. *Solid State Ion.*, **178** (13–14), 973–979.

47 Kim, S. and Park, S.J. (2007) Preparation and electrochemical behaviors of polymeric composite electrolytes containing mesoporous silicate fillers. *Electrochim. Acta*, **52** (11), 3477–3484.

48 Kim, S., Hwang, E.J. and Park, S.J. (2008) An experimental study on the effect of mesoporous silica addition on ion conductivity of poly(ethylene oxide) electrolytes. *Current Appl. Phys.*, **8** (6), 729–731.

49 Dominey, L.A. (1992) *Extended Abstracts of the Annual Automotive Technology Development Contractors' Coordination Meeting*, vol. 2, Dearborn, Michigan, pp. 2–5.

50 Ryoo, R. and Kim, J.M. (1995) Structural order in MCM-41 controlled by shifting silicate polymerization equilibrium. *J. Chem. Soc. Chem. Commun.*, **7**, 711–712.

51 Sreekanth, T., Reddy, M.J., Subramnyam, S., and Subba Roa, U.V. (1999) Ion conducting polymer electrolyte films based on (PEO+KNO$_3$) system and its application as an electrochemical cell. *Mater. Sci. Eng. B*, **64** (2), 107–112.

52 Reddy, M.J. and Chu, P.P. (2002) Optical microscopy and conductivity of poly(ethylene oxide) complexed with KI salt. *Electrochim. Acta*, **47** (8), 1189–1196.

53 Fan, L., Nan, C.W. and Zhao, S. (2003) Effect of modified SiO$_2$ on the properties of PEO-based polymer electrolytes. *Solid State Ion.*, **164** (1–2), 81–86.

54 Xiong, H.M., Zhao, K.K., Zhao, X., Wang, Y.W., and Chen, J.S. (2003) Elucidating the conductivity enhancement effect of nano-sized SnO$_2$ fillers in the hybrid polymer electrolyte PEO–

SnO$_2$–LiClO$_4$. *Solid State Ion.*, **159** (1–2), 89–95.

55 Evans, J., Vincent, C.A., and Bruce, P.G. (1987) Electrochemical measurement of transference numbers in polymer electrolytes. *Polymer*, **28** (13), 2324–2328.

56 Druger, S.D., Nitzam, A., and Ratner, M.A. (1983) Dynamic bond percolation theory: a microscopic model for diffusion in dynamically disordered systems. I. Definition and one-dimensional case. *J. Chem. Phys.*, **79** (6), 3133–3142.

57 Druger, S.D., Nitzam, A., and Ratner, M.A. (1985) Generalized hopping model for frequency-dependent transport in a dynamically disordered medium, with applications to polymer solid electrolytes. *Phys. Rev. B*, **31** (6), 3939–3947.

58 Croce, F., Curini, R., Martinelli, A., Persi, L., Ronci, F., Scrosati, B., and Caminiti, R. (1999) Physical and chemical properties of nanocomposite polymer electrolytes. *J. Phys. Chem. B*, **103** (48), 10632–10638.

59 Xi, J. and Tang, X. (2004) Nanocomposite polymer electrolyte based on poly(ethylene oxide) and solid super acid for lithium polymer battery. *Chem. Phys. Lett.*, **393** (1–3), 271–276.

60 Chen, H.W., and Chang, F.C. (2001) The novel polymer electrolyte nanocomposite composed of poly(ethylene oxide), lithium triflate and mineral clay. *Polymer*, **42** (24), 9763–9769.

61 Kim, S. and Park, S.J. (2009) Interlayer spacing effect of alkylammonium-modified montmorillonite on conducting and mechanical behaviors of polymer composite electrolytes. *J. Colloid Interface Sci.*, **332** (1), 145–150.

62 Choi, B.K. and Kim, Y.W. (2004) Conductivity relaxation in the PEO–salt polymer electrolytes. *Electrochim. Acta*, **49** (14), 2307–2313.

63 Kim, J.M., Kwak, J.H., Jun, S.N., and Ryoo, R. (1995) Ion-exchange and thermal-stability of MCM-41. *J. Phys. Chem.*, **99** (45), 16742–16747.

64 Chu, P.P., Reddy, M.J., and Kao, H.M. (2003) Novel composite polymer electrolyte comprising mesoporous structured SiO$_2$ and PEO/Li. *Solid State Ion.*, **156** (1–2), 141–153.

65 Xi, J., Qiu, C., Zhu, W., and Tang, X. (2006) Enhanced electrochemical properties of poly(ethylene oxide)-based composite polymer electrolyte with ordered mesoporous materials for lithium polymer battery. *Micropo. Mesopo. Mater.*, **88** (1–3), 1–7.

66 Kao, H.M., Tsai, Y.Y., and Chao, S.W. (2005) Functionalized mesoporous silica MCM-41 in poly(ethylene oxide)-based polymer electrolytes: NMR and conductivity studies. *Solid State Ion.*, **176** (13–14), 1261–1270.

8
Mechanics of Materials for Li-Battery Systems

Katerina E. Aifantis, Kurt Maute, Martin L. Dunn, and Stephen A. Hackney

8.1
Introduction

In the previous chapters, it was shown that not only materials selection but also materials design plays an important role in the resulting electrochemical performance of rechargeable Li-battery electrodes. The underlying physical mechanism that allows the design of the electrode to be of importance is that the mechanical response of materials is highly dependent on configuration. By understanding, therefore, the mechanical response of various electrode configurations, it will be possible to predict the most stable design that will allow for a greater lifetime since, as was shown in the previous chapters, fracture is one of the limiting factors concerning capacity retention in the next-generation electrodes.

Although there exist myriad studies and literature on the experimental aspects of Li electrodes, there is limited work that focuses on the mechanical issues of such systems. This chapter highlights some mechanical considerations that arise in the use of Li-ion batteries, including stresses that arise during manufacturing, packaging, and service (Figure 8.1). Therefore, in the sequel, after describing explicitly to role of mechanics in battery cells, an overview is given of studies that perform stress and fracture analysis on anodes and cathodes comprising of active/ inactive materials. As described in Chapter 6, in order to avoid damage and eventual fracture induced by the 300% volume expansions that Si and Sn experience on electrochemical cycling, the battery community has embarked on the exploration of nanoscale structures and composites, but with very limited understanding of the underlying mechanics processes accompanying electrochemical cycling. The need for the development of a robust theoretical material mechanics framework has already been pointed out by workers in the field; see, for example, the quotation of Beaulieu *et al.* [1]: "In the lithium alloys studied here, enormous strain can be caused with zero applied stress. The strain is caused by the incorporation of interstitial Li atoms between the existing M atoms of the alloys. It is our opinion that the theories of elasticity in solids are not suited to describe the colossal volume changes described here. We invite theorists to take up the challenge to describe these phenomena."

High Energy Density Lithium Batteries. Edited by Katerina E. Aifantis, Stephen A. Hackney, and R. Vasant Kumar
© 2010 WILEY-VCH Verlag GmbH & Co. KGaA, Weinheim
ISBN: 978-3-527-32407-1

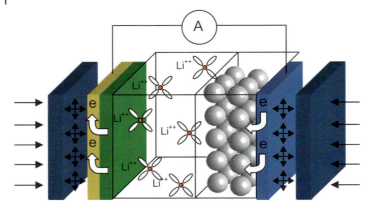

Figure 8.1   Schematic of a structurally enclosed lithium battery.

An initial theoretical consideration of Li-battery anodes was done by Wolfenstine [2], who predicted that the critical size of fully lithiated Sn and Si that will result in no fracture is below their unit cell size. This was concluded using linear elasticity, which, however, may not be valid at such small length scales. Further theoretical considerations were developed by Huggins and Nix in 2001 [3a] who performed a fracture analysis for metallic thin film anodes. They developed a simple one-dimensional decrepitation model for capacity loss predicting a terminal particle size of the order of a fraction of a micron. This model borrowed heavily from earlier work concerned with internal stress development and fracture [3b] of an epitaxial thin film on a compliant substrate. Therefore, it was not applicable to radial geometries for the active nanoparticles considered, which as shown in the previous chapters appear to be the most promising for the next-generation Li electrodes.

The mechanics challenges that arise for electrodes comprised of active nanoparticles encased/embedded in a matrix, described in Chapter 6, were undertaken by Aifantis *et al*. [4–7], through the development of simple physically based linear elasticity and fracture mechanics models with the goal of establishing preliminary easy-to-use material/component design criteria for nanostructured anodes [7] with optimum capacity, strength, and electrochemical stability. In particular, under continuous electrochemical cycling, (nano) cracking takes place at the particle–matrix interface and a (nano) damage zone is developed as documented by transmission electron microscopy (TEM) observations [4, 8]. On assuming that this damage zone is comprised by a number of radial cracks, it was possible to derive [5, 6] stability criteria for crack growth depending on the geometry assumed for the active nanoparticles (spheres, fibers/tubes, or disks/platelets) and the associated boundary conditions. It should be emphasized that the above mentioned simple mechanics models were directly motivated by the nanoscale configurations overviewed in Chapter 6, and will be described in the following section.

Strain considerations for radial geometries for the active sites have also been recently examined in [9, 10] by also employing elasticity considerations for moderately expanding cathodes, similar to those advanced previously in [4–7]. Moreover, Doyle *et al.* [9, 10] presented an interesting analysis incorporating also the effect of stress-assisted diffusion accompanying electrochemical cycling (see also the earlier work in [11]). In any case, all these mechanical models should be considered as being rather qualitative since accurate measurements of the elastic and fracture properties were not available and the underlying damage morphology was not known.

A more recent study by Golmon *et al.* [12] highlights how stresses are built up in porous electrodes that are an agglomerate of active particles, the electrolyte, and inactive material during operation. As this phenomenon occurs at multiple length scales, these scales will be described, in the sequel, along with an approach to model the interaction between electrochemical and mechanical phenomena in Li-ion battery systems. The multiple scales consist of the battery system, the porous electrode, and the individual particles in the electrode. The modeling, analysis, and simulation approaches at each scale will be described according to [12]. Mainly, the efforts involve the coupling of diffusion and mechanical behavior, along with a review of the state-of-the-art in modeling fracture and degradation of electrode particles.

8.2
Mechanics Considerations During Battery Life

Like most manufactured products, components in Li batteries are subjected to mechanical loads from their manufacture throughout their lifecycle. These mechanical loads, which occur at multiple scales, arise from various physical processes and influence the yield, performance, and ultimately the failure of Li batteries. From a macroscopic perspective, the packaging of a battery influences its ability to withstand mechanical loads during service. Different cell geometries are routinely fabricated to meet manufacturing, cost, and performance criteria; these include cylindrical, button, prismatic, and pouch cells, as were described in Chapter 1. While cylinders and buttons offer the best mechanical ruggedness due to their protective metal can, most rechargeable Li-ion batteries are fabricated in prismatic or pouch cells (Figure 8.2). These are lighter than cylindrical cells due to the absence of the metal can but offer the flexibility to design shapes and sizes to precise application-specific dimensions and also offer the best use of available space. For example, nearly every generation of laptop or cell phone has a new battery architecture. However, they are not as mechanically rugged as cylindrical cells and must be designed to prevent or accommodate expansion that can occur as pressure builds up during gas generation on charging and discharging. Pouch cells, packaged in flexible foil materials, are also more susceptible to external mechanical loads, during operation such as tension, compression, bending, and twisting. They are particularly sensitive to highly localized forces that can puncture the packaging. On the positive side, they are the most promising existing battery architecture/packaging concept for integration into multifunctional structures such as load-bearing composite materials [12–16].

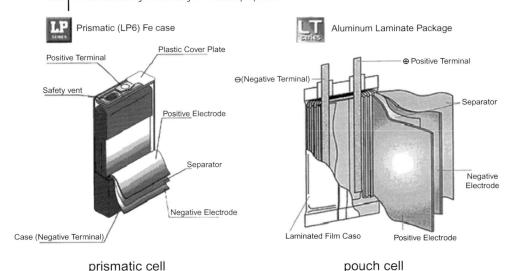

prismatic cell   pouch cell

Figure 8.2 Schematics showing construction of prismatic and pouch cells [16b].

At the heart of most current manufacturing processes is the synthesis of appropriate electrode materials and their integration in a largely planar form including an electrolyte, current collectors, and packaging. Before, however, looking into porous electrodes, it is vital to understand fracture in the next-generation anodes since, as was shown in Chapter 6, it is this fracture that inhibits the commercialization of these high-capacity materials for porous electrodes. The next-generation anodes were described to have the form of nanocomposites and nanowires, and hence, appropriate theoretical models for treating mechanical stability in such systems will also be presented. For example, in [17], significant performance enhancements were achieved through the use of silicon nanowire anodes.

During operation (discharging), Li^+ ions are extracted from the anode, transported through the electrolyte, and inserted into the cathode. During charge, the opposite occurs. We will describe details of the physical processes more thoroughly in our subsequent discussions, but in order to highlight the mechanical issues involved, and their sources, we focus here on the extraction and insertion of Li^+ ions into a cathode consisting of active $Li_yMn_2O_4$ particles; similar fracture mechanisms occur for other Li-active materials.

Insertion and extraction of Li^+ ions into a $Li_yMn_2O_4$ particle is a diffusive process driven by electrochemical reactions on the surface of the particle. As $Li_yMn_2O_4$ is reversibly cycled between $y = 0.2$ and 1, the lattice distorts with a 6.5% volume change [19]. Recent studies, however, document that the volume expansion for $LiMn_2O_4$ cathodes is 14% and results from the phase transformation that occurs upon Li insertion. In particular, Figures 8.3 and 8.4 depict $LiMn_2O_4$ electrode particles exhibiting large strain levels and damage before and after electrochemical cycling [4]. On charging, $LiMn_2O_4$ experiences a 14% volume contraction due to

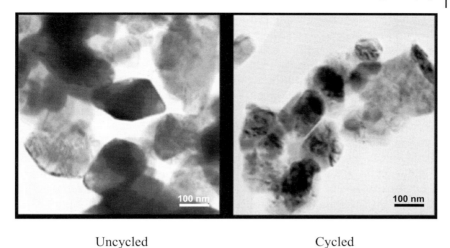

Uncycled   Cycled

Figure 8.3 Parent Mn_2O_4 and cycled $LiMn_2O_4$ samples. After multiple charge–discharge cycles, the strain within the crystals is evident from the high-frequency spatial variation of contrast in the TEM image; taken from [4].

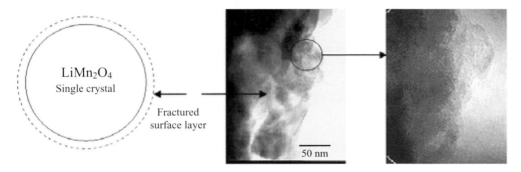

Figure 8.4 Deep discharged $LiMn_2O_4$ particles with fractured surface layers; chemomechanical stresses develop as a result of Li insertion/deinsertion; taken from [4].

the crystal structure re-ordering resulting from the formation of Mn_2O_4. On deep discharge, the respective expansion that the material experienced resulted in residual strain, which is visible in Figure 8.3. After further cycling, TEM observations show a high density of nanocracks in a single Mn_2O_4 crystal as shown in Figure 8.4. Fracture of individual $Li_yMn_2O_4$ particles can also be found in [20], while fracture in other cathodic materials such as $LiCoO_2$ and $LiFePO_4$ has been documented in [21] and [22], respectively. Similar damage processes have been documented for anodic materials as described in Chapter 6. However, damage in anodic materials, such as Si and Sn, is much more severe since the volume expansions are far grater, reaching values up to 300%, on the formation of Li-rich phases.

Particle fracture has been linked to performance degradation due to loss of electrical contact between cracked particles that increases electrical resistance and an increase of particle surface area that is subjected to detrimental side reactions [23]. Large volume mismatches (16–17% for cathode materials and >100% for anode materials) exist between the coexisting phases that result in large stresses [23, 24]. If the concentration of Li$^+$ ions was uniform in the particle, it would expand and/or contract uniformly. However, the diffusive process results in Li$^+$ concentration gradients within Li$_y$Mn$_2$O$_4$ particles that develop and evolve during insertion and extraction. This results in incompatible strains and leads to stresses that vary with position through the particle.

Although we are not discussing it in detail here, these electrochemical and mechanical phenomena are also strongly coupled to heat generation [25, 26]. Zhang *et al.* [26] have recently described these issues and developed a particle-scale model that connects electrochemistry, mechanics, and heat generation for Li-ion cells. Finally, we note that our discussion has focused on individual particles. In an aggregate, additional complexities arise due to the interparticle interactions that arise upon mechanical or electrochemical-induced loading.

8.3
Modeling Elasticity and Fracture During Electrochemical Cycling

8.3.1
Fracture in a Bilayer Configuration

The first pure mechanical study that tried to predict the mechanical effects that electrochemical cycling has on battery materials was by Huggins and Nix [3]. The configuration they considered was that of a bilayer system (Figure 8.5), by which the active material of the anode (e.g., Sn) is deposited on a substrate (Cu). Their motivation was taken by the experimental evidence of [27] that showed a better capacity retention in Sn nanoparticles as their diameters decreased.

During electrochemical cycling, the Sn layer suffers a phase transformation that corresponds to a volume expansion. If the substrate were not present, the Sn layer would be able to expand freely. The Cu layer is taken to be a compliant substrate, which means that as the Sn layer expands, on charging, elastic strains are exerted

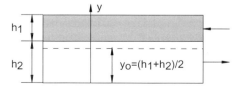

Figure 8.5 Configuration considered in [2], taken from [2].

onto the Cu, leading to the development of internal stresses and possibly fracture at the interface. The dilation that the Sn layer would experience, on Li-ion insertion, in the absence of constraints can be written as $e_T = \Delta V/V$, which implies that for the configuration at hand, the in-plane stress-free misfit strain at the bilayer interface is given by

$$\varepsilon_o = \frac{e_T}{3} \tag{8.1}$$

In order for the two layers to remain mechanically compatible, during the expansion of Sn, the Sn film experiences tensile stresses to remain flat, while Cu experiences compressive stresses. By considering strain compatibility at the interface and stress equilibrium, the stresses required for the layers to remain flat are calculated [3] to be

$$\sigma_1^{\text{flat}} = -B\varepsilon_o \frac{h_2}{h_1 + h_2}$$
$$\sigma_2^{\text{flat}} = B\varepsilon_o \frac{h_1}{h_1 + h_2} \tag{8.2}$$

where B is the biaxial elastic modulus of the material. Next the average stress that develops at the bottom layer when the system is allowed to bend freely is calculated by considering the bending moments of a plate

$$\overline{\sigma}_2^{\text{bend}} = -\frac{3B\varepsilon_o h_1^2 h_2}{(h_1 + h_2)^3} \tag{8.3}$$

The overall stress that the Cu substrate undergoes is therefore

$$\overline{\sigma}_2 = \sigma_2^{\text{flat}} + \overline{\sigma}_2^{\text{bend}} = \overline{\sigma}_2 = B\varepsilon_o \frac{h_1(h_1^2 - h_1 h_2 + h_2^2)}{(h_1 + h_2)^3} \Rightarrow \overline{\sigma}_2 = B\varepsilon_o(1-\alpha)(1-3\alpha+3\alpha^2) \tag{8.4}$$

where $h = h_1 + h_2$ and $h_2 = \alpha h$. After having a prediction for the stress in the Cu layer, Huggins and Nix [3] continued with trying to estimate the stress at which the system will fracture. On the basis of the Griffith–Orwin criterion, they calculated that the stress at which the lower layer will fracture is [3]

$$\sigma_{\text{fracture}} = \frac{K_{I_c}}{\sqrt{\pi h_2}} = \frac{K_{I_c}}{\sqrt{\pi h}} \frac{1}{\sqrt{\alpha}} \tag{8.5}$$

where K_{I_c} is the fracture toughness of Cu. Therefore, fracture will take place when the stress exerted on the Cu, due to the expansion of the Sn, exceeds the fracture stress

$$\frac{\overline{\sigma}_2}{\sigma_f} > 1 \tag{8.6}$$

Inserting Eqs. (8.4) and (8.5) in Eq. (8.6) and allowing the estimation of the parameter α provided in [3], the critical film thickness below which fracture cannot occur is calculated as [3]:

$$h_c \approx \frac{23}{\pi}\left(\frac{K_{I_c}}{B\varepsilon_o}\right)^2 = \frac{23}{\pi}\left(\frac{3K_{I_c}}{Be_T}\right)^2 \tag{8.7}$$

It was not possible to develop precise design criteria as the K_{I_c} for lithiated anodic materials is not known. Allowing, however, for the maximum and lower values of K_{I_c}, it is predicted by Eq. (8.7) that the critical film size below which fracture will not occur is 0.035 microns for very brittle materials, while for ductile materials, it is 8 microns.

8.3.2
Elasticity and Fracture in an Axially Symmetric Configuration

The recent experimental results that were presented in Chapter 6, suggest that the configuration that allows for the highest capacities and life cycle of Li anodes is that of active metals (such as Si or Sn) embedded or surrounded by carbon. On the basis of the experimental evidence of Figure 8.4, it can be seen that the surface of active sites crumbles, on cycling; this damage is shown schematically in Figure 8.6. This section focuses therefore on axially symmetric nanocomposites comprised of a highly active material (Sn or Si) with respect to Li and a less active matrix (carbon, soda glass).

Aifantis and Hackney [4] were the first who performed a mechanics study on axially symmetric anodes. The schematic configuration they considered is shown in Figure 8.7a. As a first step, they used plane stress conditions to look at the purely elastic response of the unit cell on maximum Li insertion; this plane stress condition essentially corresponds to a thin film (active/less active nanostructured) anode, which, however, has no effect from the substrate. They used this "thin film" consideration motivated by the initial experiments shown in Chapter 6. In continuing this analysis, they inserted a damage zone at the active site–matrix interface, which accounts for the fracture that takes place as shown in Figures 8.4 and 8.6. Since the matrix is more brittle from the metal active site, it is assumed that the

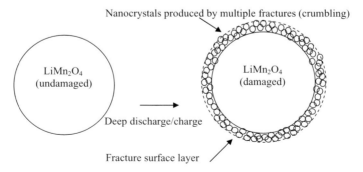

Figure 8.6 Schematical representation of fracture zone in a single crystal LiMn$_2$O$_4$ due to nanocrack formation (damaged surface layer).

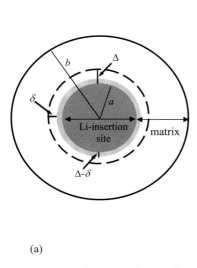

(a)   (b)

Figure 8.7 Configuration of unit cell used in analysis; a and b are the radii of the active site and matrix, respectively, Δ is the free expansion of the active site if it were not constrained (by the matrix), δ is the radial distance the matrix pushes back as it opposes this expansion, and ρ is the crack radius. (a) Corresponds to the elastic response of the unit cell; taken from [4], (b) corresponds to the unit cell configuration upon crack formation; taken from [5].

damage zone forms inside the matrix, as shown in Figure 8.7b. Since this region is severely damaged, it supports only radial stresses, and can therefore be approximated by a number of radial cracks with length ρ-a.

In modeling the purely elastic response of nanocomposite anodes, the configuration of Figure 8.7a is employed. The state of maximum Li insertion is considered and therefore the stress inside the active is constant and does not vary with r; for this state, linear elasticity in polar coordinates gives the stress and displacement expressions in the active site as

$$\sigma_r = \sigma_\theta = 2C_s; u_r = 2C_s(1 - v_s)r/E_s \tag{8.8}$$

where C_s is an integration constant, v_s is the Poisson's ratio in the active site and E_s is the elastic modulus of the active site. In the matrix, the stresses and displacement vary with r since they decrease as the distance from the active particle increases

$$\sigma_r = \frac{A_m}{r^2} + 2C_m, \sigma_\theta = -\frac{A_m}{r^2} + 2C_m, u_r = \frac{1}{E_m}\left[-\frac{A_m(1 - v_m)}{r}\right] + 2C_m(1 - v_m)r \tag{8.9}$$

where A_m and C_m are integration constants, v_m is the matrix Poisson's ratio, and E_m is the matrix elastic modulus.

To obtain the complete expressions, the constants C_s, A_m, and C_m were solved from the following boundary conditions:

Since conditions of plane stress are assumed, this maximum stress-free volumetric expansion of the Sn corresponds to an increase of its radius, denoted by Δ; therefore, the initial radius of this particle on maximum Li insertion would be: $r = a + \Delta$. The surrounding matrix, which is present under the given confined configuration, however, opposes the aforementioned free expansion by pushing back the particle by a distance $-\delta$; therefore, the final radius of the active site is $r = a + \Delta - \delta$. Thus, the total displacement of the outer surface of the particle at $r = a + \Delta$ is $u_r = -\delta$.

As the inner surface of the matrix annulus and the outer surface of the particle are in contact, it follows that once the Sn reaches its maximum expansion Δ, the matrix pushes back a distance $-\delta$, and thus, the final inner radius of the matrix is $r = a + \Delta - \delta$. Therefore, the first displacement condition for determining the constants inside the matrix is given at the interface ($r = a$) as $u_r(a) = \Delta - \delta$.

For the formulation of the outer boundary condition, at $r = b$, two different cases are considered at the outer annulus surface: (i) The first case corresponds to the "manufacturing condition" suggested by the most common configuration, according to which the whole battery system is tightly constrained by the outer casing. Therefore, the displacements at the boundary of adjacent unit cells cancel each other, and the condition at $r = b$ is taken to be $u_r = 0$. (ii) The second case corresponds to having the single particles dispersed so far apart so that they do not constrain each other. In that case the outer surface is unconstrained and the stress there is zero. This condition is referred as a "natural" condition, since the pressure that is induced on the matrix by the active site fades with increasing distance and hence the external pressure is zero. Thus, the stress exerted by the active site on the matrix is fades away with increasing distance and becomes vanishingly small at the outer cell boundary, so $\sigma_r = 0$ at $r = b$.

Consideration of the above boundary conditions allows the constants in the stress and displacement expressions to be computed as

$$C_s = -\frac{E_s \delta}{2(1 - v_s)(a + \Delta)} \tag{8.10}$$

$$A_m = -\frac{E_m a b^2 (\Delta - \delta)}{(b^2 - a^2)(1 + v_m)} ; C_m = -\frac{E_m a (\Delta - \delta)}{2(b^2 - a^2)(1 + v_m)} \tag{8.11a}$$

for the "manufacturing consistent case", while for the stress-free case, the constants in the matrix are

$$A_m = -\frac{E_m a b^2 (\Delta - \delta)}{b^2 (1 + v_m) + a^2 (1 - v_m)} ; C_m = -\frac{E_m a (\Delta - \delta)}{2[b^2 (1 + v_m) + a^2 (1 - v_m)]} \tag{8.11b}$$

It is now left to define Δ and δ. Δ is related to the free expansion on maximum Li insertion, which is 300% (for this plane stress case $\pi(a + \Delta)^2 = 3$). The distance δ can be calculated through strain energy considerations as follows. For a radial configuration under plane stress, the strain energy per unit volume and the strain energy per unit length are given, respectively, as

$$w = \frac{1}{2}\sigma_{ij}\varepsilon_{ij} = \frac{1}{2E}(\sigma_r^2 + \sigma_\theta^2 - 2\nu\sigma_r\sigma_\theta), \text{ and } W = \int_A w dA = 2\pi \int_r w r dr \qquad (8.12)$$

Combining appropriately Eqs. (8.8)–(8.11) with Eq. (8.12) allows the determination of the strain energy per unit length in the active site and matrix. The total strain energy of the system at hand is simply that of the active site plus that of the matrix $W_{tot} = W_g + W_s$. By setting the derivative equal to zero = 0 and solving for δ, we obtain for the manufacturing consistent case:

$$\delta = -\frac{E_m\Delta[a^2(1+\nu_m)+b^2(1-\nu_m)](a+\Delta)^2(1-\nu_s)}{E_s a^2(a^2-b^2)(1-\nu_m^2)-E_m[a^2(1+\nu_m)+b^2(1-\nu_m)](a+\Delta)^2(1-\nu_s)} \qquad (8.13a)$$

and for the stress-free case

$$\delta = -\frac{E_m\Delta(b^2-a^2)(a+\Delta)^2(1-\nu_s)}{E_s a^2[a^2(1-\nu_m)-b^2(1+\nu_m)]-E_m(b^2-a^2)(a+\Delta)^2(1-\nu_s)} \qquad (8.13b)$$

Now that all parameters have been defined, the stress profile inside the unit cell is shown in Figure 8.8. As the material parameters are not precisely known, this is just a qualitative plot.

The similarity of the plots obtained for cases (i) and (ii) implies that consideration of the system under "manufacturing consistent" or "natural" conditions does

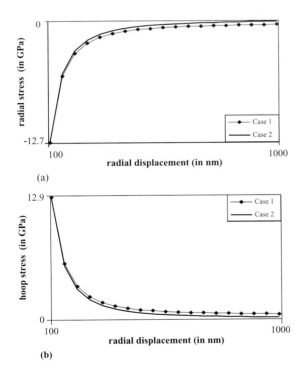

(a)

(b)

Figure 8.8 Stress distribution in matrix for purely elastic response of unit cell (a) radial stress and (b) hoop stress; taken from [4].

not affect the stress distribution inside the electrodes for this purely elastic idealized case.

8.3.3
Fracture and Damage Evolution for Thin Film Case

After modeling the elastic response, through the introduction of the quantities Δ and δ, the evolution of facture, which is responsible for capacity fade, was set forth to consider. Therefore, the damage zone of Figure 8.7b will be accounted for in the sequel [5], resulting in three regions that must be considered within a mechanics framework: the active particle, the cracked matrix, and the uncracked matrix. In considering fracture, geometry must be accounted for, since it has been observed that similar materials give different capacities depending on their configuration.

Li *et al.* [28], for example, fabricated SnO using the same method, but in one case, it had the form of a thin film, while in the other (through use of a nanoporous template), it had a nanofiber structure. As described in Chapter 6, the nanofiber structure had a capacity retention of 790 mAh/g for 800 cycles, while the thin film had a starting capacity of 580 mAh/g that decreased to 400 mAh/g after 50 cycles. The authors attributed the optimum cyclability of the nanofiber SnO configuration to the ability of this structure to better accommodate the volume expansions. Both the thin film and the fiber configurations will be considered in the sequel.

In continuing the elasticity problem, conditions of plane stress will initially be adopted, where the active site can be, for example Sn, while the matrix Li_2O, according to [28]. The stress distribution in the active site remains the same as it were in the Section 8.3.2.1, as again the state of maximum Li insertion is accounted for. Since the stress exerted by the active site on the matrix is constant and compresses the matrix, $\sigma_r = 2C_s = -p$, where p is the internal pressure. This allows the displacement boundary condition at the active site/matrix to take the form

$$u_r(a) = \Delta - \delta = \Delta - \frac{p(1-v_s)(a+\Delta)}{E_s} \tag{8.14}$$

Next, the stresses and displacement conditions inside the damage region, $a \leq r \leq \rho$, must be accounted for. Within the damage region, the stress equilibrium relation is written [5] so as to support only radial stresses as

$$\frac{d\sigma_r}{dr} + \frac{\sigma_r}{r} = 0 \Rightarrow \sigma_r(r) = \frac{k}{r} \quad \text{for} \quad a \leq r \leq \rho \tag{8.15}$$

where k is an integration constant. Given that $\sigma_r(a) = -p$, it follows that $k = -pa$. Furthermore, Hooke's law gives

$$\sigma_r = E_m \frac{du_r}{dr} \Rightarrow \frac{du_r}{dr} = -\frac{pa}{E_m r} \Rightarrow u_r(r) = -\frac{pa}{E_m}\ln(r) + u^*, \quad \text{for} \quad a \leq r \leq \rho \tag{8.16}$$

where an expression for the constant u^* can be found by setting the displacement, $u_r(\rho)$, right in front of the crack tip (i.e., just inside the uncracked region) equal to a constant $u_+(\rho)$ [29].

$$u_r(\rho) = u_+(\rho) \Rightarrow u_+(\rho) = -\frac{pa}{E_m}\ln(\rho) + u^* \Rightarrow u^* = \frac{pa}{E_m}\ln(\rho) + u_+(\rho) \qquad (8.17)$$

Then insertion of Eq. (8.17) in Eq. (8.16) concludes that inside the cracked region, the radial displacement is given by

$$u_r(r) = \frac{pa}{E_g}\ln\left(\frac{\rho}{r}\right) + u_+(\rho), \quad \text{for} \quad a \le r \le \rho \qquad (8.18)$$

The uncracked region can be thought of as a disc annulus, whose internal pressure is equal to that present at the interface with the fractured region (i.e., at $r = \rho$) and it can be found by direct substitution in Eq. (8.15):

$$p_* = -\sigma_r(\rho) = \frac{pa}{\rho} \qquad (8.19)$$

The displacement solution for a plane strain configuration (i.e., a hollow cylinder) that is subjected to an internal and external pressure is given in [30]; for the present case, it is modified to plane stress as

$$u_+(r) = \frac{r}{2\mu_m}\left\{ p_* \frac{\dfrac{b^2}{r^2} + \dfrac{1-v_m}{1+v_m}}{\dfrac{b^2}{\rho^2} - 1} - q \frac{\dfrac{\rho^2}{r^2} + \dfrac{1-v_m}{1+v_m}}{1 - \dfrac{\rho^2}{b^2}} \right\}, \quad \text{for} \quad \rho \le r \le b \qquad (8.20)$$

where $\mu_m = E_m/[2(1 + v_m)]$ is the shear modulus of the uncracked matrix and $u_+(r)$ denotes the corresponding displacement inside this region.

Thus, by setting $r = \rho$ in Eq. (8.20), the following expression for $u_+(\rho)$ is obtained:

$$u_+(\rho) = \frac{\rho[b^2(p_* + p_* v_m - 2q) - \rho^2 p_*(v_m - 1)]}{2\mu_m(v_m + 1)(b^2 - \rho^2)} \qquad (8.21)$$

Now, insertion of Eq. (8.21) in Eq. (8.18) gives a second expression for the displacement at $r = a$:

$$u_r(a) = \frac{pa}{E_g}\left\{ \ln\left(\frac{\rho}{a}\right) + 2\frac{b^2 - Cb\rho}{b^2 - \rho^2} - (1 - v_g) \right\} \qquad (8.22)$$

where $C = qb/pa$; while the displacement at $r = b$ can be deduced from Eq. (8.20) as

$$u_r(b) = \frac{pa}{E_m}\left\{ 2\frac{b\rho - Cb^2}{b^2 - \rho^2} + C(1 + v_m) \right\} \qquad (8.23)$$

Stability Index Formulation

For the present configuration, the hoop stress, σ_θ, is the opening tensile stress responsible for crack stability and growth. In [29, 31] it was shown that for such a system, this hoop stress is given as

$$\sigma_\theta(\rho^+) = \frac{pa}{b}\left\{\frac{1+(\rho/b)[(\rho/b)-2C]}{(\rho/b)[1-(\rho/b)^2]}\right\} \tag{8.24}$$

while the corresponding energy release rate for such a configuration is given as [29, 31]

$$G(\rho) = \frac{\pi\rho}{E_g n}\sigma_\theta^2(\rho^+) \tag{8.25}$$

Finally, the stability index can be defined according to [29, 31] as

$$\kappa = \frac{b}{G}\frac{dG}{d\rho} \tag{8.26}$$

Boundary Conditions   It can be seen that the energy release rate depends on the material parameters (E, v), the geometric parameters (a, b), the number of cracks n, as well as on the internal and external pressures p and q. The internal pressure can be found by equating Eqs. (8.14) and (8.22) and solving for p, while the external pressure q need not be defined explicitly as long as C is. Below are solutions for three different cases; the first two of which are in accordance to those examined in the previous Section 8.3.2.

Case 1: Clamped outer boundary (tightly constrained case)

As was explained before for the manufacturing constistent case, $u(b) = 0$ and in view of Eq. (8.23), one can solve for C to obtain

$$C = \frac{2b\rho}{(1-v_m)b^2+(1+v_m)\rho^2}$$

p is solved as explained before through Eqs. (8.14) and (8.22), and therefore, Eqs. (8.24) and (8.25) give

$$G_1(\rho) = \frac{\pi p_1^2 a^2}{nE_m\rho}\left[\frac{(1-v_m)b^2-(1+v_m)\rho^2}{(1-v_m)b^2+(1+v_m)\rho^2}\right]^2 \tag{8.27a}$$

where

$$p_1 = \Delta\left\{\frac{a}{E_m}\left[\ln\left(\frac{\rho}{a}\right)+2\left(\frac{1-v_m}{1+v_m}\right)\frac{b^2}{(1-v_m)b^2+(1+v_m)}-\frac{1}{1-v_m}\right]+\frac{(a+\Delta)}{\Gamma}\right\}^{-1}$$

It should be noted that $\Gamma = E_s/(1-v_s)$.

Case 2: Natural boundary condition (stress-free case)

If the natural boundary condition is considered, the stress at the outer boundary is zero; $q = 0$ (which implies that $C = 0$), and through Eqs. (8.14) and (8.22) p is found. Then, by using Eqs. (8.24) and (8.25), the energy release rate, G, is readily calculated as

$$G_2(\rho) = \frac{\pi p_2^2 a^2}{nE_m\rho}\left[\frac{b^2+\rho^2}{b^2-\rho^2}\right]^2 \tag{8.27b}$$

where

$$p_2 = \Delta \left\{ \frac{a}{E_m} \left[\ln\left(\frac{\rho}{a}\right) + 2\frac{b^2}{b^2 - \rho^2} - (1 - v_m) \right] + \frac{(a + \Delta)}{\Gamma} \right\}^{-1}$$

Case 3: Self-equilibrated loading

The final case to be considered, which was not accounted for in Section 8.3.2, is that of "self-equilibrated loading", that is, the force (qb) that is exerted on the matrix annulus by the surrounding unit cell is equal to that (pa) exerted onto it by the Li-insertion site. Therefore, $C = 1$, and solving for the internal pressure as before, the energy release rate, G_3 is found as

$$G_3(\rho) = \frac{\pi p_3^2 a^2}{n E_g \rho} \left(\frac{b - \rho}{b + \rho} \right)^2 \tag{8.27c}$$

where

$$p_3 = \Delta \left\{ \frac{a}{E_g} \left[\ln\left(\frac{\rho}{a}\right) + 2\frac{b}{b + \rho} - (1 - v_g) \right] + \frac{(a + \Delta)}{\Gamma} \right\}^{-1}$$

8.3.4
Fracture and Damage in Fiber-Like/Nanowire Electrodes

A more promising fabrication method that was described in the Chapter 6 is that of anodes that are fabricated as long fiber-like cylinders. The mechanical analysis for this case follows directly from that of Section 8.3.3. If the configuration of the active sites is similar to that of Figure 8.7b but with the active sites being cylindrical, it resembles that of active fibers being embedded within a matrix. In that case conditions of plane strain exist. The formulation of Section 8.3.3, which is for plane stress, can be modified to plane strain by allowing for a reduced elastic modulus and Poisson's ratio. Therefore, E and v in all the equations of Section 8.3.3 should be replaced by E' and v' [5], that is,

$$E \rightarrow E' = \frac{E}{1 - v^2}, v \rightarrow v' = \frac{v}{1 - v}$$

8.3.5
Spherical Active Sites

Another configurations described in the Chapter 6 is that of spherical active sites encased in carbon. Therefore, a similar analysis to the above was performed for spherical symmetry. The steps are the same as in Section 8.3.3, except that now the stresses in the active site are written for spherical symmetry as

$$\sigma_r = \sigma_\theta = \sigma_\phi = 2\frac{1 + v_s}{1 - 2v_s} D_s; u_r = \frac{2(1 + v_s)}{E_s} D_s \tag{8.28}$$

And, hence, the boundary condition at the active site–matrix interface takes the form

$$u_r(a) = \Delta - \delta = \Delta - \frac{(1-2v_s)(\Delta+a)p}{E_s} \tag{8.29}$$

The stress equibrium relation inside the severely fractured damage zone is written for this configuration as

$$\frac{d\sigma_r}{dr} + \frac{2\sigma_r}{r} = 0 \Rightarrow \sigma_r(r) = \frac{k}{r^2}, \quad \text{for} \quad a \leq r \leq \rho \tag{8.30}$$

which implies the displacement expression

$$\sigma_r(r) = E_m \frac{du_r}{dr} \Rightarrow \frac{du_r}{dr} = -\frac{pa^2}{E_m r^2} \Rightarrow u_r(r) = \frac{pa^2}{E_m r} + u^*, \quad \text{for} \quad a \leq r \leq \rho \tag{8.31}$$

where the constant of integration k is $k = -pa^2$; u^* is found similarly as before by setting the displacement right in front of the crack tip $u_+(\rho)$ equal to $u_r(\rho)$, that is,

$$u_+(\rho) = u_r(\rho) = \frac{pa^2}{E_m \rho} + u^* \Rightarrow u^* = u_+(\rho) - \frac{pa^2}{E_m \rho} \tag{8.32}$$

The displacement expression, therefore, inside the damage zone can be written as

$$u_r(r) = \frac{pa^2}{E_m}\left(\frac{1}{r} - \frac{1}{\rho}\right) + u_+(\rho) \quad \text{for} \quad a \leq r \leq \rho \tag{8.33}$$

Finally, the displacement expression inside the uncracked matrix, that is, a hollow sphere which experiences an internal pressure (by the active site) and an external pressure (by the neighboring unit cell), is given in [30] as

$$u_+(r) = \frac{(1+v_m)r}{E_m}\left[p_* \frac{b^3/(2r^3) + (1-2v_m)(1+v_m)}{b^3/\rho^3 - 1} \right.$$
$$\left. -q\frac{\rho^3/(2r^3) + (1-2v_m)(1+v_m)}{1-\rho^3/b^3}\right], \quad \text{for} \quad \rho \leq r \leq b \tag{8.34}$$

The internal pressure p_* is the pressure exerted at $r = \rho$, and is $p_* = pa^2/\rho^2$.

It follows that the by letting $r = \rho$ in Eq. (8.34), an analytical expression can be obtained for $u_+(\rho)$, which can then be substituted in Eq. (8.33) for the development of a second boundary condition at the active site–matrix interface as

$$u(a) = \frac{pa^2}{E_m}\left\{\frac{1}{a} - \frac{2(1-v_m)}{\rho} + \frac{3(1-v_m)(b^3 - Sb\rho^2)}{2\rho(b^3 - \rho^3)}\right\} \tag{8.35}$$

where, $S = qb^2/(pa^2)$. Now, the displacement at the outer boundary ($r = b$) is found by direct substitution in Eq. (8.34):

$$u(b) = \frac{pa^2}{2bE_m}\left\{\frac{3(1-v_m)(b^2\rho - Sb^3)}{b^3 - \rho^3} + S(1+v_m)\right\} \tag{8.36}$$

In concluding, the hoop stress and the energy release rate during crack growth for a spherical system has been calculated to be [31]

$$\sigma_\vartheta(\rho^+) = \frac{pa^2}{b^2}\left[\frac{1-3S(\rho/b)^2+2(\rho/b)^3}{2(\rho/b)^2\left(1-(\rho/b)^3\right)}\right] \tag{8.37}$$

$$G(\rho) = \frac{2(1-v_m)\rho\sigma_\theta^2(\rho^+)}{nE_m} \tag{8.38}$$

Boundary Conditions   Following the reasoning as in the thin film case, the following expressions [5] are obtained for the three boundary conditions at hand

Case 1: Clamped outer boundary (tightly constrained case)

For the clamped outer boundary $u(b) = 0$, Eq. (8.36) is set equal to zero and, therefore, S is found to be

$$S = \frac{3(1-v_m)b^2\rho}{2(1-2v_m)b^3+(1+v_m)\rho^3}$$

and hence

$$G_1 = \frac{2(1-v_m)a^4p_1^2}{nE_m\rho^3}\left[\frac{(1-2v_m)b^3-(1+v_m)\rho^3}{2(1-2v_m)b^3+(1+v_m)\rho^3}\right]^2 \tag{8.39a}$$

where

$$p_1 = \Delta\left\{\frac{a^2}{\rho E_m}\left[\frac{\rho}{a}-2(1-v_g)+\frac{3(1-v_m)(1-2v_m)b^3}{2(1-2v_m)b^3+(1+v_m)\rho^3}\right]+\frac{(a+\Delta)}{\Gamma^{sph}}\right\}^{-1}, \text{ and}$$

$$\Gamma^{sph} = \frac{E_s}{1-2v_s}$$

Case 2: Natural boundary condition (stress-free case)

Letting $q = 0$ (i.e. $S = 0$), provides

$$G_2 = \frac{(1-v_m)a^4p_2^2}{2nE_m\rho^3}\left[\frac{b^3-2\rho^3}{b^3-\rho^3}\right]^2 \tag{8.39b}$$

where

$$p_2 = \Delta\left\{\frac{a^2}{\rho E_m}\left[\frac{\rho}{a}-2(1-v_m)+\frac{3(1-v_m)b^3}{2(b^3-\rho^3)}\right]+\frac{(a+\Delta)}{\Gamma^{sph}}\right\}^{-1}$$

Case 3: Self-equilibrating loading

Finally, for the self-equilibrating case, $qb^2 = pa^2 \Rightarrow S = 1$, and therefore

$$G_3 = \frac{(1-v_m)a^4p_3^2}{2nE_m\rho^3}\left[\frac{(b+2\rho)(b-\rho)}{b^2+b\rho+\rho^2}\right]^2 \tag{8.39c}$$

where

$$p_3 = \Delta \left\{ \frac{a^2}{\rho E_m} \left[\frac{\rho}{a} - \frac{(1 - v_m)}{2} - \frac{3(1 - v_m)\rho^2}{2(b^2 + b\rho + \rho^2)} \right] + \frac{(a + \Delta)}{\Gamma^{sph}} \right\}^{-1}$$

8.3.6
Stability Plots

In order to obtain a qualitative comparison between the aforementioned theoretical model and experimental data, the stability index for the various configurations and boundary conditions is plotted in Figure 8.9 [5]; it should be noted that the stability index for all configurations is given by Eq. (8.26). The stability index indicates the energy difference (final–initial) required for crack growth to take place, therefore the more negative κ, the more difficult it is to initiate crack growth, and hence the growth is stable. Once κ becomes positive, as in the stress-free case, crack growth becomes unstable and complete fracture of the electrode can take place rapidly. An asymptote, as in the tightly constrained case, indicates that the cracks will close shut before reaching the distance at which the asymptote is observed.

Comparing therefore the plots of Figure 8.9, it can be seen that the natural boundary condition (Case 2) is the least stable since it allows for positive values of the stability index, predicting therefore unstable growth and complete fracture

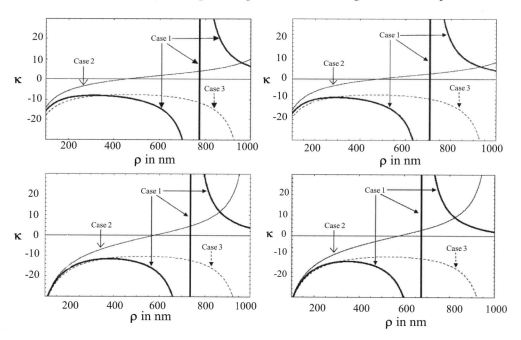

Figure 8.9 (a) Stability of thin film anode, (b) stability of fiber-like anode, (c) stability of spherical active sites embedded in glass matrix, (d) stability of spherical sites embedded in Fe–C matrix (*b* and *a* are kept the same for all cases); taken from [5].

of the electrode once cracking initiates. The asymptote observed for the tightly constrained case suggests that for Case 1, the cracks will shut once a critical distance is attained, while for the self-equilibrated loading case, stable growth is predicted until the outer boundary is reached. In particular, based on these stability plots, it can be seen that for Case 2 unstable growth will take place as the ρ-intercept is reached, approximately at $\rho_c = 450$ nm (thin film), 500 nm (fiber), 560 nm (spherical sites in glass matrix), 600 nm (spherical sites in Fe–C matrix). Larger values for ρ_c in this stress-free case imply that the anode can be used for a longer period before cracking becomes unstable and hence complete fracture takes place, disabling the anode. On the basis of this criterion, because of its better mechanical stability, the fiber-like configuration will allow for a grater capacity retention over a greater period of cycles, than the thin film configuration (as fracture occurs, the capacity decreases as shown in Chapter 6). This is a first theoretical interpretation of the superior cyclability of SnO nanofibers as compared with SnO thin films that was observed in [28] and described in detail in Chapter 6. Furthermore, based on these ρ_c considerations, the spherical Sn embedded in the Fe–C matrix is predicted to be the most preferable, since it allows for a greater crack radius (ρ_c) to be achieved before growth becomes unstable. Similar conclusions to the above can be drawn also by comparing the distance ρ at which the asymptote for Case 1 occurred. It should be noted that for the tightly constrained boundary condition, it is desired to have small distances at which the asymptote occurs, since this suggests that the distance that the cracks will propagate before they close shut will be smaller, and therefore mechanical damage will be minimized. On the basis of this criterion, again the spherical configuration is the most desirable, followed by the fiber-like inclusions and then the thin film case.

In addition to configuration, material selection also plays a significant role in the capacity of anodes. Therefore, in [5], two cases were considered for the spherical Sn sites, embedded in a matrix. In one case, the matrix was soda glass (Figure 8.9c), while in the other, it was Fe–C. It can be seen from Figure 8.9c,d that the Fe–C matrix is predicted to allow for greater stability since ρ_c was the larger for Case 1, but smaller for Case 2 (as compared to the soda glass matrix). Furthermore, in [6], the case where Si was embedded in soda glass was considered. The stability graphs were very similar to those in which Sn was active site, indicating that the formulation at hand could not distinguish between active site materials, as it could for matrix materials.

8.3.7
Volume Fraction and Particle Size Considerations

In addition to materials selection and configuration, particle size and volume fractions of active sites have been shown to affect the capacity of materials, and this can be traced back to the fact that particle size and volume fractions affect the mechanical integrity of nanocomposites.

In making volume fractions considerations, only the spherical configurations was examined, since it exhibited the best mechanical stability; therefore, Eqs. (8.37) and (8.38) were employed [7].

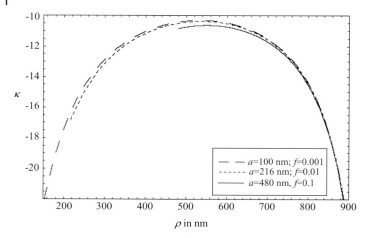

Figure 8.10 Stability plots for systems with different active site volume fractions; taken from [7].

8.3.7.1 Information from Stability Index

In order to determine the size of the active sites that will result in a more stable battery system, we plot the stability index κ with respect to the crack radius ρ, for $b = 1\,\mu m$ and varying a. The various stability behaviors obtained for Case 3 are shown in Figure 8.10. The active sites are taken to comprise of Si, while the matrix is of soda glass [7].

The more negative the values at which the stability curves start out, the more stable the system, since a greater energy difference is required for crack growth to initiate. It can therefore be predicted from Figure 8.10 that smaller active sites (i.e., smaller active site volume fractions f, where $f = a^3/b^3$) result in more stable anodes, since crack growth is initiated with greater difficulty for such systems.

8.3.7.2 Griffith's Criterion

On the basis of Griffith's theory, a crack will continue growing as long as the energy released (G) during its propagation is greater from the energy (G_c) required to create the new crack surface. By plotting therefore Eq. (8.38) together with the fracture energy (G_c) of the matrix material, which is a material constant, we can determine the crack radius at which the two energies intersect, and hence, estimate the distance at which crack growth will stop.

Aifantis et al. [7] plotted G for various active site volume fractions together with G_c. The active sites were taken to be Si, whereas the matrix was Y_2O_3. It can be seen that the smaller the volume fraction of the active sites, the smaller the distance at which crack growth stops. Hence, it is again predicted that smaller volume fractions of the active sites are more stable since they allow for less cracking to occur. Furthermore, by plotting G and G_c, for Si active sites, but different matrix materials, and constant volume fractions (i.e., $a = 100\,nm$, $b = 1\,\mu m$), we can predict which matrix material allows for the smallest crack propagation distance before cracking ceases; the results are shown in Table 8.1, where it can be seen that the

Table 8.1 Distance at which cracking stops for various matrix materials when $a = 100$ nm and $b = 1$ μm; taken from [7].

Material	G_c (J/m²)	Critical crack radius ρ at which cracking stops (nm)
SrF$_2$	0.36	725
ThO$_2$	2.5	450
Y$_2$O$_3$	4.6	425
KCl	0.14	840

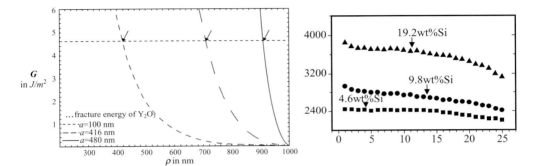

Figure 8.11 (a) Griffith's criterion for various volume fractions (b is kept 1000 μm), of Si nanospheres (with radius a) embedded in a matrix; smaller Si volume fractions crack less; taken from [7]. (b) Experimental data exhibiting the capacity fade of nano-Si embedded in sol–gel graphite matrix (taken from [33]); lower Si content has a better capacity fade. Theory and experiment are in agreement.

most preferable matrix material, based on these considerations, is Y$_2$O$_3$ (that is why it as used in Figure 8.11). It is noted that in the construction of Figure 8.11, the number of radial cracks, n, present had to be assumed, as Eq. (8.38) suggests. It should be emphasized here that example of matrix materials are taken from compounds with well-known mechanical constants. Not all of these materials have been examined for suitability in Li-battery applications (i.e. KCl), but do provide a benchmark for other materials selection protocols.

As seen in Figure 8.11a, Griffith's criterion predicts that smaller active site volume fractions allow for smaller crack distances to be attained before cracking stops. This indicates that low volume fractions will yield better capacity retention on cycling, since the cracking that leads to capacity fade is minimized. This prediction is in fact in accordance with experimental observations [33] that concluded that nanocomposites with lower Si content embedded in a sol–gel–graphite matrix had a significantly higher capacity retention (Figure 8.11b), than the corresponding higher Si content anodes (more information on this material was given in Chapter 6). Hence, despite the straightforward mechanical formulation, it is in accordance with experimental data.

In order to more precisely apply the aforementioned mechanics formulations, respective mechanical experiments must be performed to obtain not only the actual mechanical properties of nanomaterials, but also the number n of cracks that form (here it is assumed that $n = 20$). It should be noted that even if the actual elastic modulus value for nano-Si was employed in this section, the qualitative patterns would remain the same, that is, smaller active sites are more stable and hence crack growth initiation is more difficult, and they also allow for smaller crack distances.

8.3.8
Critical Crack Length

In addition to the above volume fraction considerations and use of Griffith's criterion to determine when fracture will stop, direct use of the opening tensile stress (Eq. (8.37)) can provide information concerning the critical crack length at which fracture of the electrode will occur, as well as the size of the active sites that theoretically will result in no cracking. In particular, Eq. (8.37) gives us the tensile crack opening stress as a function of the crack radius (ρ), for any given material and geometric parameters. If, however, we set σ_θ of Eq. (8.37) equal to the ultimate tensile strength of the matrix, and we define all the material and geometric parameters, the only unknown we are left with is ρ. Hence, we can solve numerically for ρ to obtain the critical crack radius and corresponding critical crack length, at which fracture of the electrode will take place for the corresponding system.

In particular, it is of interest to keep the geometric parameters the same, while we vary the material parameters so that we can obtain which material selections allow for the greatest critical crack lengths. Such a sample comparison is done in Table 8.2; Si or Sn are used as the active site, hence $E_s = 112\,\text{GPa}$ and $v_s = 0.28$ or $E_s = 44\,\text{GPa}$ and $v_s = 0.33$, while various ceramics are considered as the matrix. Two limiting cases are examined: (i) $a = 10\,\text{nm}$ and $b = 1\,\mu\text{m}$, (ii) $a = 750\,\text{nm}$ and $b = 1\,\mu\text{m}$. It should be noted that the volume expansion of Si and Sn on maximum Li insertion is approximately 300%; this corresponds to $\Delta = a^{1/3} - a$.

Table 8.2 Critical crack length for various ceramic matrices; taken from [7] (material parameters estimated from *CRC Mat Sci & Eng Handbook*).

Material	Ultimate tensile strength (MPa)	Elastic modulus (GPa)	Poisson's ratio	Critical crack Radius ρ (nm) for active sites when $a = 10\,\text{nm}$		Critical crack Radius ρ (nm) for active sites when $a = 750\,\text{nm}$	
				Si	Sn	Si	Sn
Al_2O_3	255	345	0.23	93	76	650.9	649.5
B_4C	155	450	0.21	123	99	651.6	650.9
BeO	246	400	0.24	97	78	651	649.8
WC	345	700	0.24	87	69	650.4	648.8
ZrO_2	175	140	0.23	95	81	651.3	650.4

Table 8.3 Active site size that results in no cracking when $b = 1\,\mu m$ [7].

Material	Radius (a) (nm) of active site that results in no cracking when $b = 1\,\mu m$
Al_2O_3	757
B_4C	781
BeO	746
WC	745
ZrO_2	758

Unfortunately, the same materials could not be used for both Tables 8.1 and 8.2 since for the materials in Table 8.1, the ultimate tensile strength could be found in the literature, while for the materials in Table 8.2, the critical fracture energy (G_c) could not be found. It can be seen from Table 8.2 that B_4C is the most preferable matrix materials since it allows for the greatest critical crack radius and hence more electrochemical cycles can be performed before the electrode fractures. However, it should again be emphasized that these tables provide a benchmark study as the mechanical properties of appropriate matrix materials are not known. Moreover, by comparing Si and Sn active sites, it can be seen that Si are more preferable because they always allow for a longer lifetime of the anode as they allow for greater crack lengths prior to fracture.

Furthermore, from Eq. (8.37), we can estimate the size of the active sites that will result in no cracking. When the crack radius (ρ) equals the radius of the active site (a), it implies that the crack length ($\rho - a$) is zero and hence no cracks are present. Therefore, by defining the material parameters in Eq. (8.37), where again σ_θ = ultimate tensile strength, setting $\rho = a$, and choosing a particular b, we can estimate the corresponding radius a that will theoretically result in no cracking. Such sample results are shown in Table 8.3.

8.3.9
Mechanical Stability of Sn/C Island Structure Anode

An alternative materials design that would seem to combine the fracture resistance of small particles with the electrical connectivity of thin films would be an island structure, in which the particles of electrochemically active material are physically attached to a larger substrate to produce "islands" of active materials. This configuration corresponds to the Sn/C composites that were described in Chapter 6, Section 6.7.1.3 [32, 34], which were shown to have a superior electrochemical performance as opposed to other Sn/C configurations. Their unique capacity retention is attributed to the greater mechanical stability this microstructure could maintain during the continuous volume changes experienced by the Sn on Li insertion and deinsertion. The advantage of the island configuration relative to the free standing particles or continuous thin films is proposed to be due to improvements in both electrical connectivity and mechanical stability.

When islands of active material are attached to larger particles of carbon, the electrical contact with the current collector can be facilitated as long as there exists contact between the larger carbon particles. This is significantly easier than attempting to maintain the electrical connectivity between each individual, free standing Sn nanoparticle during the volume expansion or contraction in electrochemical cycling. The mechanics-based advantage of the island design relative to the continuous thin film geometry is related to the resistance to delamination during volume expansions. This resistance to delamination of islands compared with thin films arises from two sources: the stress reduction associated with free boundary conditions and the aspect ratio. For a given strain due to island volume expansion on an infinitely stiff substrate, it is proposed that the critical delamination compressive stress concept developed in [35, 36] may be adapted to the island geometry as

$$\sigma_c = \frac{(15)E}{12(1-v^2)}\left(\frac{2H}{L}\right)^2 \tag{8.40}$$

where E is the modulus, v is Poisson's ratio, H is the island half thickness and L the island half width. That is, if the compressive stress in the island exceeds the critical delamination stress, then the island will detach and most likely fail to contribute to the electrochemical activity. Thus, it can be seen that the aspect ratio (H/L) is critical to the mechanical stability of the island geometry. In order to test the idea that the free surface boundary conditions associated with the island geometry will reduce the compressive stress relative to that in a continuous thin film, a 2D elasticity solution is pursued for the geometry shown in Figure 8.12. The stresses and displacement that the island experiences are taken to be given by the biharmonic Marguerre solutions [37] as

$$\sigma_{xx}(x,y) = \frac{E}{1-v^2}\left(-\frac{d^3\Psi}{dx^2dy}+v\frac{d^3\Psi}{dy^3}\right), \quad \sigma_{yy}(x,y) = \frac{E}{1-v^2}\left((1+v)\frac{d^3\Psi}{dx^2dy}+\frac{d^3\Psi}{dy^3}\right),$$

$$\sigma_{xy}(x,y) = \frac{E}{1-v^2}\left(\frac{d^3\Psi}{dx^3}+\frac{d^3\Psi}{dxdy^2}\right), \quad u_x(x,y) = \frac{1+v}{1-v}\left(\frac{d^2\Psi}{dxdy}\right)$$

$$\tag{8.41}$$

where the energy functional Ψ is of the form

$$\Psi = \cos(\beta_n x)(A\sinh(\beta_n y) + B\cosh(\beta_n y) + C\beta_n y\sinh(\beta_n y)) \tag{8.42}$$

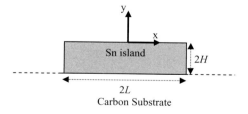

Figure 8.12 Schematical representation of Sn/C island microstructure.

where $\beta_n = n\pi/2L$ (n = odd), and the constants A and B are found from the boundary conditions

$$\sigma_{xx}(x,H) = 0, \sigma_{yy}(x,0) = 0, \sigma_{xy}(x,0) = 0 \tag{8.43}$$

as

$$A = 0 \quad \text{and} \quad B = -\frac{2vC}{1+v}$$

where C is found from the definition of the displacement at the island–carbon interface as a linear function of x so that $\Delta x/L = u_x(y = -2H)$. Expanding the displacement as a Fourier series in which C becomes a Fourier coefficient and thus a function of the Fourier index, n, the various values of C_n may be determined from Eq. (8.44):

$$\int_{-L}^{L}\left(\frac{\Delta x}{L}\sin(\beta_n x)\right)dx = \int_{-L}^{L}[u_x(-2H)\sin(\beta_n x)]dx \tag{8.44}$$

where Δ, as in the previous sections, denotes the unconstrained expansion of the Sn island.

Hence,

$$C_n = \frac{32L^3(v-1)\Delta\sin(n\pi/2)}{n^3\pi^3(n\pi)[n\pi H(1+v)\cosh(n\pi H/L) - L(v-1)\sinh(n\pi H/L)]}$$

Insertion of the constants of integration in Eq. (8.42) and then in (8.41) allows the determination of the compressive stress σ_{xx}; Δ is taken to be a proportional function of the strain associated with the maximum constraint to island expansion, ε, $\Delta = \varepsilon L$. The form of the spatial variation in the normalized compressive stress along x at $y = (-H)$ for two different H/L values at a given island volume is shown in Figure 8.13. It may be seen that the average and maximum amplitude of the

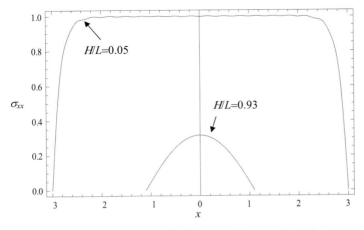

Figure 8.13 Comparing the normalized compressive stress for different H/L values.

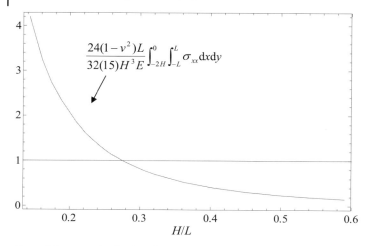

Figure 8.14 In order for the islands to remain attached to the carbon, their height/length ratio (*H/L*) must be greater than 0.3.

compressive stress is reduced in value when H/L is large as compared to when it is small.

Application of the critical stress for delamination (Eq. (8.40)) to this elasticity solution for the island structure with individual Sn island volume 16.75 nm³ is obtained by calculating the average compressive stress in an island. This average compressive stress is substituted into Eq. (8.40) giving

$$\frac{2L}{8HL^2}\int_{-2H}^{0}\int_{-L}^{L}\sigma_{xx}dxdy = \frac{15E}{12(1-v^2)}\left(\frac{2H}{L}\right)^2 \tag{8.45}$$

Rearrangement of terms results in the cancellation of elastic constants or

$$\frac{24(1-v^2)L}{32(15)H^3E}\int_{-2H}^{0}\int_{-L}^{L}\sigma_{xx}dxdy = 1 \tag{8.46}$$

The result predicts that the normalized average stress must be less than 1 in order for the particle to be attached on the substrate surface. This approach then predicts that the particle height to width ratio must be above ~0.3 (Figure 8.14) in order to prevent the island detachment from the carbon substrate. The implication here is that for a given percentage of volume change during cycling, an island structure (height/width = 1) has a greater mechanical stability than a thin film (height/width ≪ 1).

In this section, it was therefore illustrated that continuum mechanics formulations can be adapted to examine fracture and damage evolution in higher energy density storage devices such as Li batteries. Although the examples considered were for anodic materials, since they experience fracture more severely, the same formulation can be applied for cathodic intercalation materials.

Of course, the optimum goal is the development of theoretical models that can couple the electrochemical and mechanical effects of electrochemical cycling. The approaches that are headed toward this direction are elaborated on in the sequel. It will be seen, however, that they are more involved and require the use of computation methods.

8.4
Multiscale Phenomena and Considerations in Modeling

Traditionally, only a subset of the phenomena describing the behavior of a battery has been considered for modeling particular performance criteria. Most commonly, the capacity and the charge and discharge behavior are predicted considering only the transport of Li^+ ions and charge at the macro- and/or microscale [38–40]. However, the significant dependency of conduction and reaction parameters requires accounting for spatially and temporarily varying temperature fields. The degradation of the battery performance due to the failure of individual particles has been the focus of the previous sections in this chapter. This and the following sections are dedicated to multiscale phenomena and the interaction between transport processes and mechanical effects.

Resolving directly all length and timescales involved in the charge and discharge process of a battery requires modeling every single particle in the electrodes. As the size of the particle decreases, the characteristic time of the diffusion process also decreases. With the typical battery thickness for lithium batteries being around 100–200 μm and the particle size of less than 10 μm, such an approach leads to a nearly intractable model that provides limited insight and, when simulated numerically, suffers from a large computational burden. Therefore, only few studies involve direct numerical simulation approaches. For a two-dimensional battery model, a numerical framework was developed in [11] to directly predict the intercalation and mechanical response of particles within the cathode; later this framework was extended to include an intercalation anode as well [41]. Furthermore, in [42], a three-dimensional microscale model was presented that could predict the electrochemical performance of batteries with random and periodic microscale cathode layouts by modeling every cathode particle. However, stress effects, both internal and external, were not included in this model.

Assuming a separation of length and timescales between macro- and microscale processes, multiscale modeling methods provide a systematic and efficient approach to describe the interaction between micro- and macroscale phenomena (see Figure 8.15). Most commonly, some form of periodic or stochastic microscale behavior at the microscale phenomena is assumed. Using a mesoscale homogenization method that accounts for the interaction of multiple particles, microscopic parameters are averaged over a representative volume element (RVE) and introduced via constitutive models into the macroscale models. Macroscopic parameters may impose boundary conditions on the RVE problem.

A multiscale model was developed by Doyle [38, 39] based on porous electrode theory that describes the transport of Li ions and charge in the cathode. The cathode is considered to be a heterogeneous material containing an electrolyte phase as well as active and nonactive material. Using analytical volume averaging

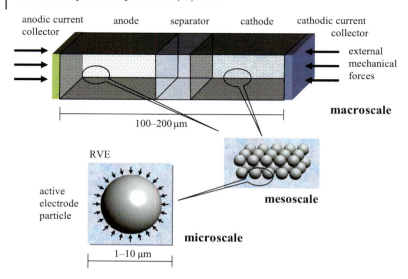

Figure 8.15 Multiscale battery model.

techniques, the macroscopic effective material parameters and an effective Li-ion absorption rate are derived as a function of the volume fraction of the active material from a semianalytical spherical particle model. Doyle's approach [38, 39] has been recently augmented by Golmon *et al.* [12] to account for mechanical interaction phenomena. This multiscale model describes the interaction between the various transport and mechanical mechanisms that are predicted to take place at the macro- and microscales. The coupling of macroscopic and microscopic strains and stresses is modeled by an implicit homogenization method based on the Mori-Tanaka model, which accounts for the interaction between particles in the host matrix. Numerical studies for liquid electrolytes based on a finite element implementation of this multiscale model showed that the stress in cathode particles strongly depends on macroscopic parameters, such as the discharge rate and external mechanical loads, and confirms previous analyses that the stress in the particles decreases with their size. In the following sections, we follow a multiscale modeling approach and focus on relevant mechanical phenomena, their interaction with Li ions and charge transport, and their impact on the battery performance at the macro- and microscales.

8.4.1
Macroscale Modeling

To simplify the following discussion, we focus on a simple layout of a lithium battery shown in Figure 8.16. Our model configuration includes an anodic current collector, a lithium-foil anode, a gel or liquid electrolyte, a porous intercalation cathode, and a cathodic current collector. The external positive and negative elec-

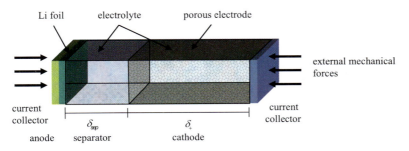

Figure 8.16 One-dimensional battery model.

tric terminals are connected to the current collectors. This model can be easily augmented for secondary (rechargeable) batteries by replacing the lithium-foil anode with an Li-intercalation anode. As the thickness of the anode-separator-cathode layer is typically much smaller than its other dimensions, a one-dimensional model is sufficient to model the electro-chemical-mechanical phenomena of interest.

As Li$^+$ ions are inserted into the cathode active material particles, the particles swell resulting in both particle- and battery-level strains and stresses. Moreover, external mechanical forces and constraints hindering the expansion of the battery also result in battery- and particle-level stresses, potentially affecting the discharge characteristic and lifetime of the battery. The performance of a battery can be characterized, among other measures, by the potential difference across the battery as a function of the utilization and discharge current. Utilization is the ratio of the actual over the maximum Li concentration that can intercalate into the active cathode material. The capacity of the battery is proportional to the area beneath the utilization. Particle stresses affect the diffusion of Li into the particles and therefore the utilization as a function of discharge current. However, numerical studies [12] have shown that this effect is, in general, negligible. Mechanical effects, on the other hand, play an important role in the performance, degradation, and failure of the batteries. In particular, as discussed in detail in Section 8.3, stress-induced fracture of active particles contributes significantly to the degradation of capacity on cycling; other effects that contribute to capacity loss, but at lower levels, are corrosion and evolution of hydrogen and oxygen gas due to side reactions [25].

The modeling of the coupled electrochemical diffusion and evolution of particle stresses will be discussed in greater detail in Section 8.5. In the following text, we briefly describe the phenomena and models at the macroscale that contribute to the generation of particle stresses and fracture at the microscale.

The macroscopic response of the one-dimensional battery configuration described above is characterized by the Li$^+$ ion concentration in the liquid phase, c_1, the electric potential of the liquid phase, ϕ_2, the electric potential of the solid phase, ϕ_1, and the macroscopic displacements, $\mathbf{u}$. The intercalation of Li$^+$ ions from the electrolyte into the particles is represented by an effective macroscopic pore wall flux, j_{eff}. The associated swelling of the particles results in a macroscopic electrochemical eigenstrain $\mathbf{e}^{ch}$. Following porous electrode theory [25], the electrochemical transport of Li$^+$ ions through the electrolyte and the current carried by the solid and liquid phases are governed by the following transport equations (Eqs. (8.47)–(8.49)) and constitutive relations (Eqs. (8.50)–(8.52)).

In particular, the transport of Li$^+$ ions through the electrolyte is due to both migration and diffusion and is given by

$$\varepsilon \frac{\partial c_1}{\partial t} - \nabla \cdot N + \frac{1}{F} \frac{\partial t_+^0}{\partial c_1} i_2 \cdot \nabla c_1 - (1 - t_+^0) j_{eff} = 0 \tag{8.47}$$

where F is Faraday's constant, $\mathbf{N}$ is the Li$^+$ ion flux, and ε denotes the volume fraction occupied by the electrolyte and is equal to one in the separator region.

It can be seen that Eq. (8.47) includes two source terms to account for the migration due to the current $\mathbf{i_2}$ carried by the electrolyte and for the effect of Li$^+$ ions leaving the electrolyte and intercalating into the solid electrode material. The transference number, t_+^0, is the percentage of the current in the solution carried by the Li$^+$ ions rather than the anions in solution; the transference number is, in general, a function of the Li ion concentration, c_l. As Li$^+$ ions leave the electrolyte and enter the solid material, this creates an effective pore wall flux, j_{eff}. The currents $\mathbf{i_1}$ and $\mathbf{i_2}$ in the solid and liquid phases are governed by Eq. (8.48) and (8.49) with source terms to account for the effects of Li entering and exiting the phases.

$$\nabla \cdot \mathbf{i_1} + F j_{eff} = 0 \tag{8.48}$$

$$\nabla \cdot \mathbf{i_2} - F j_{eff} = 0 \tag{8.49}$$

$$\mathbf{N} = D_{eff} \nabla c_1 \tag{8.50}$$

In the solid phase, Ohm's law relates the current and electric potential as

$$\mathbf{i_1} = -\lambda \nabla \phi_1 \tag{8.51}$$

where λ is the electric conductivity. In the liquid phase, the constitutive relationship is defined by a modified Ohm's law (Eq. (8.52)) that accounts for the effect of the Li$^+$ ion concentration and the temperature, T, on the current $\mathbf{i_2}$ and potential ϕ_2.

$$\mathbf{i_2} = -\kappa_{eff} \left[\nabla \phi_2 - \frac{RT}{F} (1 - t_+^0) \nabla \ln c_1 \right] \tag{8.52}$$

The universal gas constant is denoted by R. The effective electrolyte diffusivity and conductivity in the porous cathode can be computed by analytical or numerical homogenization methods. Doyle approximates the effective material properties by the Bruggeman relations [39]:.

$$D_{eff} = \varepsilon D_2 \kappa_{eff} = \varepsilon^{3/2} \kappa_\infty \tag{8.53}$$

Typically discharge and charge processes are either modeled as being current or potential controlled. At the Li-foil anode, a galvanic process model can be used to relate the current discharged to the number of lithium atoms that disassociate at the anode–separator interface. Under idealized conditions, a Li$^+$ ion enters the electrolyte for every electron that leaves the anode and all current is carried by the electrolyte. At the cathode current–collector interface, lithium cannot leave the battery, so the Li ion flux is zero, and the solid cathode material carries all the current.

To account for the coupling of chemical and mechanical phenomena at the macroscale, we extended the electrochemical model (Eqs. (8.47)–(8.49)) to include mechanical deformations:

$$\nabla \cdot \boldsymbol{\sigma} + \mathbf{b} = \mathbf{0} \qquad (8.54)$$

where $\boldsymbol{\sigma}$ is the macroscopic stress tensor and $\mathbf{b}$ the vector of body forces, which is of minor importance and will be neglected in the further discussion. Assuming small deformations and a linear elastic material behavior, we use the following linear constitutive and kinematic models:

$$\boldsymbol{\sigma} = \mathbf{C}_{\text{eff}} : (\mathbf{e} - \mathbf{e}^{\text{ch}}) \quad \mathbf{e} = \frac{1}{2}(\nabla \mathbf{u} + \nabla \mathbf{u}^{\text{T}}) \qquad (8.55)$$

where $\mathbf{C}_{\text{eff}}$ is the macroscopic elasticity tensor derived from the microstructural layout and $\mathbf{e}$ is the total macroscopic infinitesimal strain tensor due to the macroscopic displacements, $\mathbf{u}$. The effect of the swelling of particles at the microscale is modeled by the electrochemical eigenstrain tensor, $\mathbf{e}^{\text{ch}}$. Depending on the enclosure of the battery cell, the macroscopic elasticity problem is subject to one of the following two boundary conditions: either the cell is fixed on both sides preventing the cell from expanding (this corresponds to the manufacturing consistent case of Section 8.3.2) or the cell is subject to an external pressure.

The reader may note that only the diffusion of Li$^+$ ions and transport of charge in the liquid phase are explicitly coupled at the macroscale. The coupling of the other governing equations, in particular the interaction of the diffusion processes and mechanical deformations, is only through the interaction of phenomena between the macro- and microscales. The macroscopic problem depends on the effective pore wall flux, j_{eff}, and the electrochemical eigenstrain tensor $\mathbf{e}^{\text{ch}}$ which are derived from a microscale individual particle model via homogenization. In addition, the Li$^+$ ion concentration, the potential difference between the liquid and solid phases, and the macroscopic displacements impact microscale processes as discussed in the following section.

8.5
Particle Models of Coupled Diffusion and Stress Generation

In Section 8.3, we described the analysis of stress generation and fracture in particles during electrochemical cycling by noting that the cycling results in the development of a transformation strain in the Li-active particles. The development of the transformation strain is due to the diffusion of the Li$^+$ ions into and out of the electrodes during electrochemical cycling. In this section, we look more closely at models of electrochemical diffusion and the resulting stress generation.

The analysis of diffusion and stress generation in single particles is valuable, and has been pursued, from multiple perspectives. Single-particle analyses (i) serve as an idealization to describe the behavior of actual electrodes, (ii) are accurate representations of simpler electrode systems and companion-controlled experiments, for example, single particles, fibers, and microelectrodes, that are useful to help understand fundamental phenomena [46, 47] involved with individual electrode components, and (iii) describe an important scale in the multiscale modeling of electrodes and overall battery cell performance [12].

In this section we describe the analysis of a single particle, emphasizing the fundamental physical phenomena (Li$^+$ transport, electrochemical reaction kinetics, and stress generation) and the mathematical formulation of models to simulate the physical processes under idealized and realistic boundary conditions. There are various ways by which we could approach this section; for example, we could describe the general coupling between diffusion and stress in a solid [48–50] and then specialize these results to situations of relevance in batteries. It is preferred, however, and more consistent with most battery studies to focus on an individual spherical particle and describe the coupled electromechanical diffusion and stress in this context. In continuing, we then describe related connections to other shapes including more complex (ellipsoidal) and simple (planar film) ones, as well as structures that are not particle-like (structured nanowires). We step through the understanding and modeling of the physical phenomena beginning with electrochemical diffusion, then the development of stress, and finally the fracture of particles as a result of the stresses.

The studies overviewed in Section 8.3 [3, 4–7] modeled the chemical strains as eigenstrains (stress-free strains) without considering the complete connection to diffusion of Li$^+$ ions; they then calculated the resulting internal stresses and applied fracture mechanics criteria to determine critical conditions for particle fracture, as was shown. While much of our discussion and analysis is generally applicable to both cathode and anode materials, we will often refer to Li$_y$Mn$_2$O$_4$ particles (Figures 8.3 and 8.4) when it helps the discussion to be specific. The development assumes isotropic behavior, an assumption that is commonly used in the literature and justified for crystalline particles based on the fact that particles are typically aggregates of randomly oriented crystals.

8.5.1
Li$^+$ Transport During Extraction and Insertion from a Host

Consider a cathode particle, for example, Li$_y$Mn$_2$O$_4$, surrounded by an electrolyte. Extraction and insertion of Li$^+$ ions in a cathode particle can be modeled as species diffusion (Li$^+$ ions) in a solid with the surface ion flux determined by electrochemical reactions at the particle–electrolyte interface. The development described here follows that of [12, 26, 51]. Specifically, the analysis of Zhang *et al.* [51] rigorously treats the diffusion problem for a spherical particle. The study by Zhang *et al.* [26] was extended to treat the fully coupled diffusion–stress problem for a spherical particle with a traction-free surface and constant Li$^+$ ion surface flux using a well-known analogy to more common thermal stress problems. In [12], the context of treating the interactions among particles and the matrix in an aggregate is extended to include a surface traction that depends on the particle interactions in the aggregate.

The diffusion of Li^+ ions within the particle can be described by the conservation equation [26]:

$$\frac{\partial c_s}{\partial t} + \nabla \cdot \mathbf{J} = 0 \tag{8.56}$$

where c_s is the concentration and $\mathbf{J}$ is the flux of Li^+ ions. The flux $\mathbf{J}$ can be written as

$$\mathbf{J} = -c_s M \nabla \mu \tag{8.57}$$

where M is the mobility of Li^+ ions and μ is the chemical potential. Most of the existing literature that has studied the coupled diffusion–stress phenomena in electrodes has assumed an ideal solid solution (in [9, 10], both ideal and nonideal solutions were considered), so that μ is defined as

$$\mu = \mu_o + RT \ln X - \Omega \sigma_h \tag{8.58}$$

where μ_o is a constant, R the universal gas constant, T the absolute temperature, X the Li^+ ion molar fraction, Ω the Li^+ ion partial molar volume, and σ_h the hydrostatic stress, that is, $(\sigma_{11} + \sigma_{22} + \sigma_{33})/3$. For a uniform temperature and incorporating Eq. (8.58), the Li^+-ion flux can be expressed as

$$\mathbf{J} = -D_s \left(\nabla c_s - \frac{\Omega c_s}{RT} \nabla \sigma_h \right) \tag{8.59}$$

where $D_s = MRT$ is the diffusion coefficient. We take the diffusion coefficient to be independent of the stress but not a function of the state of charge of the particle. The analysis of Christenson and Newman [9, 10] appears to be the only one in the literature that incorporates the dependence of the diffusion coefficient on the state of charge in their analysis. It is interesting that the Li^+ ion flux only depends on the spatial gradient of the hydrostatic stress, $\nabla \sigma_h$, but not on the value of σ_h.

Equations (8.57) and (8.59) can be combined to describe the diffusion of Li^+ ions in the particle as

$$\frac{\partial c_s}{\partial t} = D_s \left(\nabla^2 c_s - \frac{\Omega}{RT} \nabla c_s \cdot \nabla \sigma_h - \frac{\Omega c_s}{RT} \nabla^2 \sigma_h \right) \tag{8.60}$$

This partial differential equation must be solved subject to an initial condition and boundary conditions. The initial condition is taken to be $c_s(\mathbf{x}, t = 0) = c_o$, while the diffusion boundary conditions can be of two forms:

potentiostatic (prescribed voltage)

$$c_s(\mathbf{x}, t) = C \quad \text{on } \mathbf{x} = \text{surface} \tag{8.61}$$

where C is a constant,

galvanostatic (prescribed current) boundary conditions:

$$\mathbf{J} = \mathbf{j}_s \quad \text{on } \mathbf{x} = \text{surface.} \tag{8.62}$$

Finally, the traction must be prescribed on the particle surface:

$$t_n = \sigma_{ai}\hat{\mathbf{n}} \tag{8.63}$$

where $\hat{\mathbf{n}}$ is the outward normal to the surface. Most single-particle studies have assumed that the particle surface is traction free, that is, $t_n = 0$, but Golmon *et al.* (2009) [12] incorporated the nonzero pressure boundary condition to use the single-particle solution in a homogenization approach that models the interaction among multiple particles in a porous electrode.

8.5.2
Electrochemical Reaction Kinetics

In practice, the boundary conditions on the surface of the particle are determined by the kinetics of the electrochemical reaction at the surface. In order to help understand basic phenomena, Cheng and Verbrugge [52] carried out calculations for constant potentiostatic (voltage) and galvanostatic (current) boundary conditions when there was no coupling between stress and diffusion, for example, when $\nabla \sigma_h = 0$ in Eq. (8.60). In this case, Eq. (8.61) can be solved analytically for a spherical particle. In [40], constant galvanostatic boundary conditions that incorporated the coupling between the stress and the diffusion were considered, necessitating the numerical solution of Eq. (8.60) for a spherical particle.

In [12], an approach similar to that of [9, 10, 26] was followed, and the boundary condition on the Li$^+$-ion flux was modeled using a Butler–Volmer equation that expresses the reaction kinetics at the particle surface. The reactions considered were of the type

$$\text{Li}^+ + \text{e}^- + \text{Mn}_2\text{O}_4 \underset{\text{extraction}}{\overset{\text{insertion}}{\rightleftharpoons}} \text{LiMn}_2\text{O}_4 \tag{8.64}$$

A Butler–Volmer equation gives the current density as the difference between the cathodic and anodic currents:

$$BV(c_l, \varphi_1, \varphi_2, c_s) = i_0 \left[c_s e^{\left(\frac{a_A F}{RT}(\eta - U'(c_s))\right)} - (c_T - c_s) e^{\left(-\frac{a_C F}{RT}(\eta - U'(c_s))\right)} \right] \tag{8.65}$$

where $\eta = \phi_1 - \phi_2$, while i_0 is the exchange current density:

$$i_0 = F k_2 (c_{\max} - c_1)^{\alpha_C} (c_1)^{\alpha_A} \tag{8.66}$$

The maximum Li concentration in the solid is denoted by $c_s c_{\max}$; U' is the overpotential, and k_2 is the reaction rate constant at the cathode-electrolyte interface. The transfer coefficients α_A and α_C are typically set to 0.5. The Li$^+$-ion flux at the particle surface (Eq. (8.62)) is then described by

$$BV(c_l, \phi_1, \phi_2, c_{s,\text{surf}}) - F j_s = 0 \tag{8.67}$$

8.5.3
Stress Generation

As shown in Figure 8.4, the diffusion of the Li$^+$ ions results in a lattice expansion or contraction and, therefore, the stresses that develop vary with position and time. Assuming small deformation and linear elastic behavior, the static equilibrium equation in the absence of body forces are

$$\sigma_{ij,i} = 0 (i, j = 1,2,3) \tag{8.68}$$

with the following constitutive equation:

$$\sigma_{ij} = \frac{E}{1+v} \varepsilon_{ij} + \left(\frac{Ev}{(1+v)(1-2v)} \varepsilon_{kk} - \frac{E\Omega}{2(1-2v)} c_s \right) \delta_{ij} \tag{8.69}$$

where E and v are Young's modulus and Poisson's ratio of the assumed isotropic particle, respectively. Equations (8.68) and (8.69) show that the effect of Li$^+$ ion diffusion enters the elasticity equations in the same manner as thermal expansion does. In particular, Eqs. (8.57), (8.59), (8.68), and (8.69) describe full coupling between diffusion and elasticity of the particle. Partial coupling can be described by a simplified form where Eqs. (8.68) and (8.69) remain as they are, but the terms involving $\nabla \sigma_h$ are dropped from the diffusion (Eqs. (8.57) and (8.59)). This coupling physically means that the diffusion of Li$^+$ ions affects the stress state in the particle, but stresses do not alter the diffusion characteristics. We note that in this development, only a small deformation of the particle is considered. To the best of our knowledge, this is the situation considered in all studies in the literature, concerning this coupling of phenomena, except the work presented in [9, 10]. We have implemented this formalism into a finite element approach to solve the fully coupled diffusion–stress equations in spherical coordinates [12] for a cathode material.

The authors in [9, 10] considered the finite deformation of a particle that means that the geometrical changes during deformation are sufficiently large so that the diffusion and elasticity equations must be solved at any time step on a current configuration rather than the initial, undeformed configuration. For many cathode particles with modest volume changes, especially when constrained in an aggregate, the small deformation assumption is likely adequate. It is questionable, though, when the diffusion-induced strains are large such as in silicon and tin anodes where volume changes in excess of 100% are expected.

8.5.4
Representative Results

We start by simulating the diffusion of Li$^+$ ions into a spherical Mn$_2$O$_4$ particle using the above formalism and discuss the resulting concentration and stress fields through the particle as a function of time. Similar simulation results have been presented in [52] for one-way diffusion–stress coupling and constant current or voltage boundary conditions, as well as in [9, 10] for fully coupled behavior with boundary conditions supplied by the Butler–Volmer reaction kinetics. The

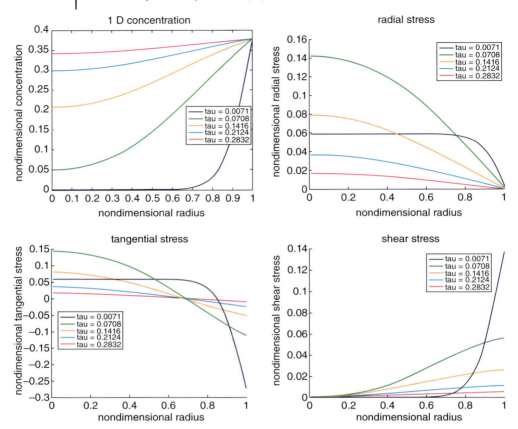

Figure 8.17 (a) Normalized concentration profile along with (b) radial, (c) tangential, and (d) shear stress profiles for simulations of Li⁺-ions insertion into a spherical Mn₂O₄ particle with the surface flux of Li⁺ ions dictated by the Butler–Volmer equation.

parameters used in the simulations of Figure 8.17 are $E = 10^9 \, \mathrm{N/m^2}$, $c_0 = 5000 \, \mathrm{mol/m^3}$, $r = 5 \, \mu\mathrm{m}$, $\eta = 4.14 \, \mathrm{V}$, $C_1 = 2000 \, \mathrm{mol/m^3}$, and $T = 300 \, \mathrm{K}$.

The development of the concentration profile with time gives rise to a radial stress that is tensile, maximum at the center of the particle, and decreases to zero at the particle surface. Over time, the radial stress increases, reaches a peak, and then decreases and becomes more uniform as the concentration profile becomes more uniform. These results for the full coupling and Butler–Volmer boundary condition are in line with [52] which show that in the case of a constant potential, the radial stress reaches its maximum value at $\tau = Dt/R^2 = 0.0574$, which shortly after the ion concentration at the center of the sphere starts to rise appreciably. The tangential stress, on the other hand, is compressive at the surface and tensile in the interior of the particle; it is equal to the radial stress at the center and the center of the particle is in a state of hydrostatic tension. The strong stress gradient decreases with time as the concentration profile becomes more uniform. The shear stress vanishes at the center of the particle and increases as the surface is approached, but unlike the radial and tangential stresses, its maximum magnitude occurs at $\tau = 0$. This behavior can be understood physically as follows. As the concentration increases in the material near the surface, its natural state is swollen with respect to the material near the center of the particle, that is, there is a mismatch in strain. This material constrains at the surface from swelling freely, which results in a tensile stress near the center of the particle. Over time as the concentration becomes more uniform, the strain mismatch between various regions of the particle decreases and so do the stresses, as well. On Li^+ ion extraction, all of these trends are reversed.

Figure 8.18 shows the simulation results for the surface flux and resulting stresses in a spherical particle during an applied potential scan of 3.454–4.138 V;

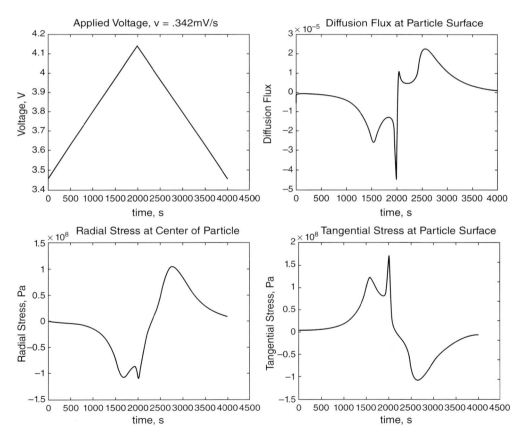

Figure 8.18 Simulations of cyclic voltametry: (a) voltage sweep, (b) surface flux of Li^+ ions, (c) radial stress at center of particle, and (d) tangential stress at surface of particle.

this corresponds to cycling the $Li_yMn_2O_4$ from $y = 0.995$ to 0.2 [26]. The potential scan rate is $0.342\,mV/s$ with $c_0 = 22{,}785.5\,mol/m^3$ and $c_i = 1000\,mol/m^3$. These results are similar to those reported in [51] for diffusion and [26] for diffusion coupled with stress; indeed an excellent discussion of the diffusion results is given by [51] and so it is not repeated here.

Figure 8.18b shows the ion flux on the particle surface due to the electrochemical kinetics for one voltammetry cycle. In this simulation, ions are extracted from the particle during the first half cycle and so the radial stress at the particle center is compressive and the tangential stress at the surface is tensile (the opposite of the results in Figure 8.17). The behavior during the voltammetry cycle is not symmetric, even though the applied potential is, as a result of the asymmetry present in the Butler–Volmer equation. Two peaks appear in the ion flux during each half cycle. In [26, 51], similar simulation results are observed (in [51], they are compared favorably to experiments) and attribute the peaks to the two plateaus that exist in the open-circuit potential of $Li_yMn_2O_4$ as a function of y. Peaks also exist in the stress versus time plots and these are attributed to significant concentration gradients that occur at the corresponding times during the diffusion process. In [26], it is shown that the nature of the response can change as a function of discharge rate, for example, the breadth of the peaks, and that higher discharge rates result in higher stresses.

Figure 8.19 shows the results of a series of simulations to illustrate the effect of the full coupling between stress and diffusion. Results are shown at $1000\,s$ for the concentration and stresses in the particle for two cases: full coupling and one-way coupling between stress and diffusion; the latter amounts to the $\nabla \sigma_h$ term vanishing in Eq. (8.59). The concentration profile illustrates that the hydrostatic stress gradient enhances diffusion as the concentration gradient through the particle decreases in the presence of stress due to the presence of the stress gradient. The stress gradient reduces the concentration gradient, which then reduces the stress and stress gradient and a balance is eventually reached [9]. In [26], it is shown that stress effectively changes the diffusion coefficient by a multiplier that depends on the hydrostatic stress gradient. The multiplier is always positive so the diffusion coefficient is effectively increased due to stress. Simulated stresses are consistent with the results presented in [9] that show that the full coupling results in reduced stress gradients of about 20% compared with the one-way coupled simulation, thus emphasizing the importance of including full coupling in simulations.

The simulations here have been done for a spherical particle and illustrate some of the features of stress development during insertion and extraction of Li^+ ions from particles in porous electrodes. The use of a similar simulation approach to treat other important issues has appeared in the recent literature and so we briefly mention some of them here. The basic formalism described consists of a full coupling between diffusion and elasticity with boundary conditions supplied by the Butler–Volmer equation that describes the electrochemical reaction kinetics. Here the basic equations have been implemented in spherical coordinates and solved with a finite element formulation. The basic equations can be solved in 2D and 3D Cartesian coordinates as well as 2D polar coordinates to simulate other

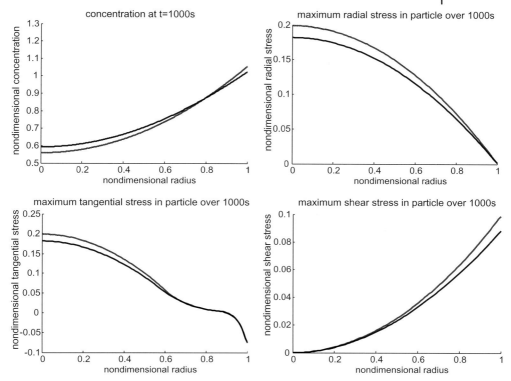

Figure 8.19 (a) Normalized concentration profile along with (b) radial, (c) tangential, and (d) shear stress profiles for simulations of Li⁺-ion insertion into a spherical Mn_2O_4 particle with the surface flux of Li⁺ ions dictated by the Butler–Volmer equation. The blue and black lines represent one- and two-way coupling between stress and diffusion, respectively.

geometrical configurations such as films and rods. In [26], this approach was used to simulate intercalation in spheroidal particles and the results show that for a prolate spheroid with an aspect ratio of about 2, the concentration is higher around the poles, the von Mises stress is highest around the equator, and the shear stress is maximum on the surface.

Perhaps the most comprehensive studies of intercalation-induced stress are those of Christensen and Newman [9, 10] who studied both carbon particles (anodes) and $Li_yMn_2O_4$ particles (cathodes). In addition to the facets of their analysis that we mentioned already, note that they also include in their work (i) the effect of a diffusion coefficient that is dependent on the state of charge, (ii) nonideal as well as ideal solutions, and (iii) the effect of phase transitions that can occur resulting in two-phase particles. Stresses that arise in multiphase particles can exceed those during the intercalation process. They also note that in a porous electrode, the current density will vary with position (see also [12] for a treatment) and this can result in a nonuniform current density at the surface of a particle that can lead to more severe stress concentrations. Finally, we note that the simulations presented here assume constant values of most material properties during intercalation, and future experimental and theoretical efforts should address the change of material properties during intercalation and deintercalation.

8.6
Diffusional Processes During Cycling

8.6.1
Multiscale Electrochemical Interactions

As discussed in Section 8.5, the microscale particle model plays a key role in understanding and predicting the macroscopic behavior of a battery. In particular, the transport equations (8.47)–(8.49) are coupled via the effective pore wall flux, j_{eff}, which is directly dependent on the microscopic particle surface flux, j_s. Using porous electrode theory, the macroscopic pore wall flux is typically computed as the average of microscopic fluxes. Assuming a uniform flux for all particles in an RVE leads to the following simple homogenization equation [39]:

$$j_{\text{eff}} = \frac{3(1-\varepsilon)}{R_s} j_s \tag{8.70}$$

Note that the effective macroscopic flux is a volume-specific quantity while the microscopic flux is related to the surface area of the particles. A broad variety of homogenization methods are available to determine the macroscopic mechanical properties as a function of the microstructural layout. In general, the effective macroscopic stress state needs to be evaluated based on the stress fields in the particles and host matrix of one RVE. Specifically, for the multiscale battery model described previously and assuming a linear elastic mechanical behavior, this task can be split up into evaluating the macroscopic material tensor $\mathbf{C}_{\text{eff}}$ and effective eigenstrain tensor $\mathbf{e}^{\text{ch}}$. In the framework of a multiscale battery model, Golmon *et al.* [12] determined $\mathbf{C}_{\text{eff}}$ and $\mathbf{e}^{\text{ch}}$ based on the Mori-Tanaka theory [43–45]. This theory is based on a particle agglomerate model and accounts for the interaction of elliptical particles within in a matrix host.

The Li intercalation and the deformations of the particle depend on macroscopic properties. As described by Butler–Volmer (Eq. (8.60)), the surface reaction kinetic is an explicit function of the macroscopic Li$^+$ ion concentration and the potential difference between the electrolyte and the particle. The interaction between the macro- and microscopic mechanical responses is more involved and highlights the importance of an accurate mesoscale homogenization model. The surface pressure acting on the particle stems from macroscale mechanical loads and the swelling of the particle due to Li intercalation. Often it is assumed that only one particle swells with an infinite host matrix. Depending on the particle density, this model may significantly underpredict the surface pressure as it ignores the swelling of neighboring particles. This effect can be accounted for via homogenization methods based on periodic or stochastically arranged RVEs. Instead of relating the surface pressure directly to the swelling of the particle at the microscale, the effective electrochemical eigenstrain, e^{ch}, is modeled by homogenizing the deformations of the particles within an RVE. The surface pressure is then formulated as a function of the macroscopic stress tensor that depends on the electrochemical eigenstrain and external loads at the macroscale. As the deformations of the particle depend on the surface pressure, this approach couples the macroscopic mechanical response and the homogenization model tightly into the microscopic particle model.

To illustrate the coupling between macro- and microscale phenomena in a battery, we present numerical results for the simulation of a single cell using the 1D battery model described previously. The battery consists of a Li-foil anode, a separator of length $\delta_{sep} = 50\,\mu m$ and a cathode of length $\delta_{sep} = 100\,\mu m$ with $\varepsilon = 30\%$ liquid volume fraction. The electrolyte is PEO-LiCF$_3$SO$_3$ and the active cathode material is TiS$_2$. We assume spherical particles with a nominal radius of $R_s = 5\,\mu m$. The influence of macro- and microscale parameters on the electrochemical and mechanical performance is studied by varying the discharge current density and the particle size. We further illustrate the influence of mechanical parameters on the battery performance by varying the mechanical boundary conditions. All simulation results are for a single current-controlled discharge process at a nominal discharge current density of $I = 10\,A/m^2$. The finite element implementation and additional studies on this battery model are given in [12].

The discharge current affects both the capacity of the battery and the stress in the particles. Figure 8.20 shows the effects of different discharge current densities

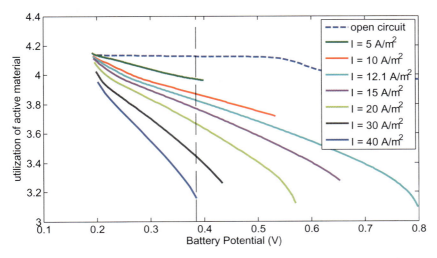

Figure 8.20 Discharge characteristics for different discharge current densities; vertical dashed line marks an average utilization in the cathode of 38.6%.

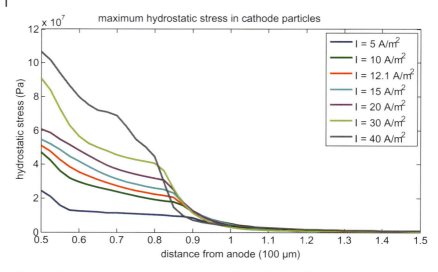

Figure 8.21 Maximum hydrostatic stress in solid particles for different discharge current during discharge densities up to 37% utilization, plotted over cathode.

on battery voltage over utilization. As expected, a higher current density leads to a lower utilization of the active material and, therefore, a lower capacity. In addition, a larger discharge current density results in a higher peak hydrostatic stress in the particles, potently shortening the lifetime of the battery. In Figure 8.21, we show the peak stress in the particles as a function of location of the particle and discharge current. The plot shows the peak stress that a particle observes during discharge to an average utilization of 37% in the cathode. This value is chosen as this is the maximum level of utilization that could be simulated for all discharge current studies (see Figure 8.20). Note we only show the cathode and ignore the separator region, as it does not comprise of active particles.

Reducing the size of the cathode particles significantly enhances the capacity and lowers the stress level in the particles. This is shown in Figures 8.22 and 8.23 by studying particle sizes between $R_s = 5 - 20\,\mu$m. These results are in agreement with numerical studies [42] and are of particular interest when considering failure due to particle fracture (see Sections 8.3 and 8.5). In Figure 8.23, we show the maximum hydrostatic stress in the particles as a function of location and size of the particle. The plot shows the peak stress a particle observes during discharge to an average utilization of 70% in the cathode.

To study the influence of the macroscopic mechanical behavior on the battery, we vary the mechanical boundary conditions. We compare the following cases: (a) we constrain the macroscopic deformation across the entire thickness of the battery, (b) we clamp the battery at the anodic and cathodic current collectors suppressing the expansion of the battery, and (c) we apply an external pressure to the current collector compressing or stretching the battery; pressure values of ±10 and

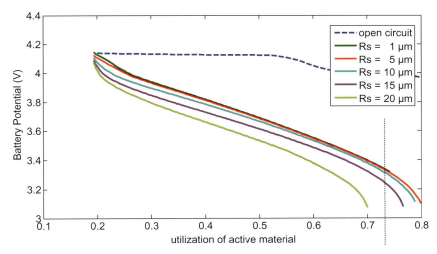

Figure 8.22 Battery potential over utilization of cathode material for different particle sizes; vertical dashed line marks an average utilization in the cathode of 70%.

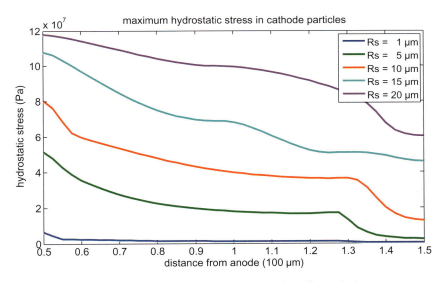

Figure 8.23 Maximum hydrostatic stress in solid particles for different discharge current during discharge densities up to 70% utilization, plotted over cathode.

±100 MPa are considered. Suppressing the macroscopic deformations in case (a) results in the surface pressure on the particle, which is only due to the homogenized electrochemical eigenstrains. Therefore, any effects due to the flexibility of the separator and the cathode are eliminated. Because of the spherical geometry of the particles, the assumption of a uniform particle surface pressure, and as the

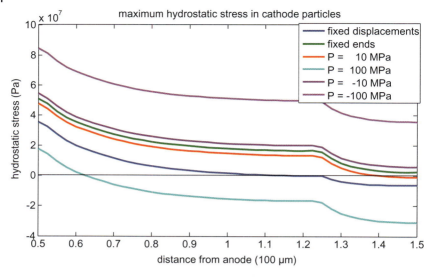

Figure 8.24 Volume average hydrostatic stress in particles plotted over cathode for different particle sizes at $t = 6540$ s corresponding to 67.27% utilization.

Li-intercalation process only depends on the stress gradient, macroscopic deformations do not affect the electrochemical performance of the battery. The intercalation in nonspherical particles varies with the particle surface pressure, and the macroscopic mechanical response will influence the capacity and discharge behavior of the battery.

However, even in the case of the spherical particles, the dependency of macro- and microscale stresses across the battery on the mechanical boundary conditions cannot be ignored, as they contribute to failure mechanisms in the battery. The maximum hydrostatic stress a particle in the cathode observes during the first 6540 s is plotted in Figure 8.24 as a function of the particle location. As is expected, greater external pressure correlates to a higher average particle stress.

8.6.2
Diffusion Stresses in Low Symmetry Composition Fields

Not all electrochemically active materials exhibit Li-concentration fields of high symmetry. For example, as considered in Chapter 5, $FePO_4$ [53] and CoO_2 [54] exhibit large anisotropy in the Li-diffusion coefficient that reduces the symmetry of the concentration field. In addition, there is evidence that the spatial variation in element compositions in some crystal systems is quasi-periodic and does not match the solutions to the linear form of Fick's second law. In these cases, the stress inside the "body" of a crystal with complicated composition variations cannot be modeled using an external load on the crystal surface, but must be rather described using the consequences of atomic forces within the "body." As such, the traditional, two-dimensional plane strain elasticity is modified to account for these body forces in a manner analogous to the Navier equation for decoupled thermal stresses. The generalized "body force" associated with the composition (lattice parameter) variation is

$$F = -Ea\nabla C \tag{8.71}$$

where α is the lattice expansion coefficient and E the elastic modulus. The body force potential, V, is related to the body force as

$$F = -\nabla V \tag{8.72}$$

The compatibility condition for plane strain relates the airy stress function φ to the body force potential as

$$\nabla^2\nabla^2\varphi = -(1-v)\nabla^2 V \tag{8.73}$$

where ∇^2 is the two-dimensional differential operator in the x- and y-spatial dimensions; the equilibrium equations for plane strain are then

$$\sigma_{xx} = \frac{\partial^2\varphi}{\partial y^2} + V, \quad \sigma_{yy} = \frac{\partial^2\varphi}{\partial x^2} + V, \quad \sigma_{xy} = \frac{\partial^2\varphi}{\partial x\partial y} + V \tag{8.74}$$

The solution to Eq. (8.73) is biharmonic with a source term. As an example, consider the case where the composition varies according to

$$C = C_o + \delta\sin\left(\frac{n\pi x}{L}\right) \tag{8.75}$$

where n is an odd integer and L is the half length of the crystal along the x- and y-directions. Equation (8.75) defines the body force and body force potential using Eqs. (8.71)–(8.73) such that

$$V = E\alpha\varepsilon\sin\left(\frac{n\pi x}{L}\right)$$

The solution to Eq. (8.73) must be obtained such that it is consistent with the stress-free surface boundary conditions. If no external forces are applied to the crystal surface, then

$$\sigma_{xx}|_{|x|=L} = 0, \quad \sigma_{yy}|_{|y|=L} = 0, \quad \sigma_{xy}|_{|y|=L} = 0$$

Given these boundary conditions, we expect the general solution of Eq. (8.73) to have the form

$$\varphi = \sin(\lambda x)(A\cosh(\lambda y) + B + Dy\sinh(\lambda y)) \tag{8.76}$$

where A, B, and D are constants to be determined through the stress-free boundary conditions and

$$\lambda = \frac{n\pi}{L}$$

Once the constants in Eq. (8.76) are determined in a manner consistent with the boundary conditions, then the stresses and strains may be evaluated through Eq. (8.74a–c).

8.7
Conclusions

Motivated by Chapter 5 and 6, this chapter illustrated how mechanics can be used to examine the materials behavior in Li-ion battery electrodes. Moreover, consideration of materials mechanics allows for the design of material structures capable of long term mechanical integrity and optimum electrochemical performance. The general conclusions that can be drawn are that smaller volume fractions of active sites allow for a greater mechanical stability and hence better capacity retention, while smaller particles allow for a higher capacity. In order to fully utilize the aforementioned theoretical considerations, it will be necessary to perform appropriate mechanistic properties characterization to determine the physical constants relevant to deformation and fracture.

References

1 Beaulieu, L.Y., Eberman, K.W., Turner, R.L., Krause, L.J., and Dahn, J.R. (2001) *Electrochem. Solid-State Lett.*, **4**, A137.

2 Wolfenstine, J. (1999) *J. Power Sources*, **79**, 111.

3a Huggins, R.A. and Nix, W.D. (2000) *Ionics*, **6**, 57.

3b Nix, W.D. (1989) *Metall. Trans.*, **20A**, 2217.

4 Aifantis, K.E., and Hackney, S.A. (2003) *J. Mech. Behav. Mater.*, **14**, 413.

5 Aifantis, K.E. and Dempsey, J.P. (2005) *J. Power Sources*, **143**, 203.

6 Aifantis, K.E., Dempsey, J.P., and Hackney, S.A. (2005) *Rev. Adv. Mater. Sci.*, **10**, 403.

7 Aifantis, K.E., Dempsey, J.P., and Hackney, S.A. (2007) *J. Power Sources*, **165**, 874.

8 Shao-Horn, Y., Hackney, S.A., Kahaian, A.J., Kepler, K.D., Skinner, E., Vaughey, J.T., and Thackeray, M.M. (1999) *J. Power Sources*, **81–82**, 496.

9 Christensen, J. and Newman, J. (2006) *J. Solid State Electrochem.*, **10**, 293.

10 Christensen, J. and Newman, J. (2006) *J. Electrochem. Soc.*, **153**, 1019.

11 Garcia, R.E., Chiang, Y.-M., Craig Carter, W., Limthongkul, P., and Bishop, C.M. (2005) *J. Electrochem. Soc.*, **152**, A255.

12 Golmon, S., Maute, K., and Dunn, M.L. (2009) Computers and Structures, **87**, 1567.

13 Thomas, J.P. and Qidwai, M.A. (2005) *J. Oper. Manage.*, **57**, 18.

14 Pereira, T., Scaffaro, R., Nieh, S., Arias, J., Guo, Z., and Hahn, H.T. (2006) *J. Micromech. Microeng.*, **16**, 2714.

15 Pereira, T., Guo, Z., Nieh, S., Arias, J., and Hahn, H.T. (2008) *Composites Sci. Tech.*, **68**, 1935.

16 (a) Pereira, T., Scaffaro, R., Guo, Z., Nieh, S., Arias, J., and Hahn, H.T. (2008) *Adv. Eng. Mater.*, **10**, 393. (b) Woodbank Communications Ltd (2005) http://www.mpoweruk.com/cell_construction.htm

17 Chan, C.K., Peng, H., Liu, G., McIlwrath, K., Zhang, X.F., Huggins, R.A., and Cui, Y. (2008) *Nat. Nanotech.*, **3**, 31.

18 Chan, C.K., Ruffo, R., Hong, S.S., Huggins, R.A., and Cui, Y. (2009) *J. Power Sources*, **189**, 34.

19 Hunter, J.C. (1981) *J. Solid State Chem.,* **39**, 142.

20 Thackeray, M.M., Johnson, C.S., Vaughey, J.T., Li, N., and Hackney, S.A. (2005) *J. Mater. Chem.*, **15**, 2257.

21 Wang, H., Jang, Y.-I.I., Huang, B., Sadoway, D.R., and Chiang, Y.-M. (1999) *J. Electrochem. Soc.*, **146**, 473.

22 Wang, D., Wu, X., Wang, Z., and Chen, L. (2005) *J. Power Sources*, **140**, 125.

23 Shao-Horn, Y., Hackney, S.A., Johnson, C.S., Kahaian, A.J., and Thackeray, M.M. (1998) *J. Solid State Chem.*, **140**, 116.

24 Amatucci, G.G., Pereira, N., Zheng, T., and Tarascon, J.-M. (2001) *J. Electrochem. Soc.*, **148**, A171.

25 Newman, J. and Thomas-Alyea, K.E. (2004) *Electrochemical Systems*, Wiley-Interscience, Hoboken.

26 Zhang, X., Sastry, A.M., and Shyy, W. (2008) *J. Electrochem. Soc.*, **155**, A542.

27 Yang, J., Winter, M., and Besenhard, J.O. (1996) *Solid State Ion.*, **90**, 281.

28 Li, N., Martin, C.R., and Scorsati, B. (2001) *J. Power Sources*, **97**, 240.

29 Dempsey, J.P., Palmer, A.C., and Sodhi, D.S. (2001) *Eng. Fract. Mech.*, **68**, 1961.

30 Westergard, H.M. (1953) *Theory of Elasticity and Plasticity*, Harvard University Press.

31 Dempsey, J.P., Slepyan, L.I., and Shekhtman, I.I. (1995) *Int. J. Fract.*, **73**, 223.

32 Sarakonsri, T., Aifantis, K.E., and Hackney, S.A. (2010) *Nanostruct. Malts*, In print.

33 Niua, J. and Yang Lee, J. (2002) *Electrochem. Solid-State Lett.*, **5**, A107.

34 Aifantis, K.E., Brutti, S., Sarakonsri, T., Hackney, S.A., and Scrosati, B. (2010) *Electrochim. Acta*, In print.

35 Marshall, D.B. and Evans, A.G. (1984) *J. Appl. Phys.*, **56**, 2632.

36 Evans, A.G. and Hutchinson, J.W. (1984) *Int. J. Solids Struct.*, **20**, 455.

37 Little, R.W. (1973) *Theory of Elasticity*, Prentice-Hall, Englewood Cliffs, NJ.

38 Doyle, M., Fuller, T.F., and Newman, J. (1993) *J. Electrochem. Soc.*, **140**, 1526.

39 Doyle, M. (1995) Design and Simulation of Lithium Rechargeable Batteries, Ph.D. Thesis. Department of Chemical Engineering, University of California, Berkeley.

40 Zhang, X., Shyy, W., and Sastry, A.M. (2007) *J. Electrochem. Soc.*, **154**, 910.

41 Garcia, R.E. and Chiang, Y. (2007) *J. Electrochem. Soc.*, **154**, 856.

42 Wang, C. and Sastry, A.M. (2007) *J. Electrochem. Soc.*, **154**, 1035.

43 Mori, T. and Tanaka, K. (1973) *Acta Metallurgica*, **21**, 571.

44 Benveniste, Y. (1987) *Mech. Mater.*, **6**, 147.

45 Benveniste, Y. and Dvorak, G.J. (1990) On a correspondence between mechanical and thermal effects in two-phase composites, in *Micromechancis and Inhomogeneity* (eds G.J. Weng, M. Taya and H. Abe), Springer, New York, pp. 65–81.

46 Verbrugge, M.W. and Kocj, B.J. (1996) *J. Electrochem. Soc.*, **143**, 24.

47 Verbrugge, M.W. and Kocj, B.J. (1996) *J. Electrochem. Soc.*, **143**, 600.

48 Prussin, S. (1961) *J. Appl. Phys.*, **32**, 1876.

49 Li, J.C.M. (1978) *Metall. Trans. A*, **9A**, 1353.

50 Yang, F. (2005) *Mater. Sci. Eng. A*, **409**, 153.

51 Zhang, D., Popov, B.N., and White, R.E. (2000) *J. Electrochem. Soc.*, **147**, 831.

52 Cheng, Y.-T. and Verbrugge, M.W. (2009) *J. Power Sources*, **190**, 453.

53 Li, J., Yao, W., Martin, S., and Vaknin, D. (2008) *Solid State Ion.*, **179**, 2016.

54 Iriyama, Y., Inaba, M., Abe, T., and Ogumi, Z. (2001) *J. Power Sources*, **94**, 175.

Index

High Energy Density Lithium Batteries. Edited by Katerina E. Aifantis, Stephen A. Hackney, and
R. Vasant Kumar
© 2010 WILEY-VCH Verlag GmbH & Co. KGaA, Weinheim
ISBN: 978-3-527-32407-1